THE PAN DICTIONAR

Carol Gibson BSc, former math
try, now devotes herself to writing scientific books on a full-
time basis.

THE PAN DICTIONARY OF

MATHEMATICS

EDITED BY CAROL GIBSON

Pan Books
London, Sydney and Auckland

First published in Great Britain 1981 by
Facts on File Publications, Oxford
This edition published 1990 by Pan Books Limited,
Cavaye Place, London SW10 9PG

9 8 7 6 5 4 3 2 1

© Charles Letts & Co Ltd 1981

ISBN 0 330 31455 6

Printed in England by Clays Ltd, St Ives plc

PREFACE

This dictionary is one of a series designed for use in schools. It is intended for students of mathematics, but we hope that it will also be helpful to other science students, and to anyone interested in the life sciences. The other books in the series are the Pan dictionaries of *Biology*, *Physics* and *Chemistry*.

We would like to thank all the people who have co-operated in producing this book. A list of writers and editors is given on the acknowledgments page. We are also grateful to the many people who have given additional help and advice.

ACKNOWLEDGMENTS

Editor
    Carol Gibson B.Sc.

Consultant
    Norman Cunliffe B.Sc.

Contributors
    Eric Deeson M.Sc., F.C.P., F.R.A.S.
    Claire Farmer B.Sc.
    Jane Farrill Southern B.Sc., M.Sc.
    Valerie Illingworth B.Sc., M.Phil.
    Alan Isaacs B.Sc., Ph.D.
    Sarah Mitchell B.A.
    Roger Adams B.Sc.
    Roger Picken B.Sc.
    Janet Triggs B.Sc.

A

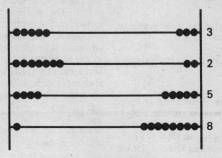

An abacus showing the number 3258 on the right hand side.

abacus A calculating device consisting of rows of beads strung on wire and mounted in a frame. An abacus with nine beads in each row can be used for counting in ordinary arithmetic. The lowest wire counts the digits, 1, 2, ... 9, the next tens, 10, 20, ... 90, the next hundreds, 100, 200, ... 900, and so on. The number 342, for example, would be counted out by, starting with all the beads on the right, pushing three beads to the left hand side of the bottom row, four to the left of the second row, and two to the left of the third row. Abaci, of various types, are still used for calculating in some countries; experts with them can perform calculations very rapidly.

Abelian group (commutative group) A type of group in which the elements can also be related to each other in pairs by a commutative operation. For example, if the operation is multiplication and the elements are rational numbers, then the set is an Abelian group because for any two elements a and b, $a \times b = b \times a$, and all three numbers, a, b, and $a \times b$ are elements in the set. All cyclic groups are Abelian groups. *See also* cyclic group, group.

abscissa The horizontal or x-coordinate in a two-dimensional rectangular Cartesian coordinate system. *See* Cartesian coordinates.

absolute Denoting a number or measurement that does not depend on a standard reference value. For example, absolute density is measured in kilograms per cubic metre but relative density is the ratio of density to that of a standard density (i.e. the density of a reference substance under standard conditions). *Compare* relative.

absolute convergence The convergence of the sum of the absolute values of terms in a series of positive and negative terms. For example, the series:
$$1 - (1/2)^2 + (1/3)^3 - (1/4)^4 + \ldots$$
is absolutely convergent because
$$1 + (1/2)^2 + (1/3)^3 + (1/4)^4 + \ldots$$
is also convergent. A series that is convergent but has a divergent series of absolute values is *conditionally convergent*. For example,
$$1 - 1/2 + 1/3 - 1/4 + \ldots$$
is conditionally convergent because
$$1 + 1/2 + 1/3 + 1/4 + \ldots$$
is divergent. *See also* convergent series.

absolute error The difference between the measured value of a quantity and its true value. *Compare* relative error. *See also* error.

absolute maximum *See* maximum point.

absolute minimum *See* minimum point.

absolute value The modulus of a real number or of a complex number. For example, the absolute value of -2.3, written $|-2.3|$, is 2.3. The absolute value of a complex number is also the modulus, for example the absolute value of $2 + 3i$ is $\sqrt{(2^2 + 3^2)}$. *See also* modulus.

abstract number A number regarded simply as a number, without reference to any material objects or specific examples. For example, the number 'three' when it does not refer to three objects, quantities, etc., but simply to the abstract concept of 'three'.

acceleration Symbol: a The rate of change of speed or velocity with respect to time. The SI unit is the metre per second per second ($m\,s^{-2}$). A body moving in a straight line with increasing speed has a positive acceleration. A body moving in a curved path with uniform (constant) speed also has an acceleration, since the velocity (a vector depending on direction) is changing. In the case of motion in a circle the acceleration is v^2/r directed to the centre of the circle (radius r).
For constant acceleration:
$$a = (v_2 - v_1)/t$$
v_1 is the speed or velocity when timing starts; v_2 is the speed or velocity after time t. (This is one of the equations of motion.) This equation above gives the mean acceleration over the time interval t. If the acceleration is not constant
$$a = dv/dt, \text{ or } d^2x/dt^2$$
See also Newton's laws of motion.

acceleration due to gravity *See* acceleration of free fall.

acceleration of free fall (acceleration due to gravity) Symbol: g The constant acceleration of a mass falling freely (without friction) in the Earth's gravitational field. The acceleration is towards the surface of the Earth. g is a measure of gravitational field strength – the force on unit mass. The force on a mass m is its weight W, where $W = mg$.
The value of g varies with distance from the Earth's surface. Near the surface it is just under 10 metres per second per second ($9.806\,65$ $m\,s^{-2}$ is the standard value). It varies with latitude, in part because the Earth is not perfectly spherical (it is flattened near the poles).

acceptance region When considering a hypothesis, the sample space is divided into two regions – the *acceptance region* and the *rejection region* (or *critical region*). The acceptance region is the one in which the sample must lie if the hypothesis is to be accepted.

accumulation point (cluster point) For a given set S, a point that can be approached arbitrarily closely by members of that set. Another way of saying this is that an accumulation point is the limit of a sequence of points in the set. An accumulation point of a set need not necessarily be a member of the set itself, although it can be. For example, any rational number is an accumulation point of the set of rationals. But 0 is an accumulation point of the set $\{1, \frac{1}{2}, \frac{1}{4}, \frac{1}{8}, \dots\}$ although it is not itself a member of the set.

accuracy The number of significant figures in a number representing a measurement or value of a quantity. If a length is written as 2.314 metres, then it is normally assumed that all of the four figures are meaningful, and that the length has been measured to the nearest millimetre. It is incorrect to write a number to a precision of, for example, four significant figures when the accuracy of the value is only three significant figures, unless the error in the estimate is indicated. For example, 2.310 ± 0.005 metres is equivalent to 2.31 metres.

acre A unit of area equal to 4840 square yards. It is equivalent to 0.404 68 hectare.

action An out-dated term for *force. See* reaction.

action at a distance An effect in which one body affects another through space with no apparent contact or transfer between them. *See* field.

actuary An expert in statistics who calculates insurance risks and relates them to the premiums to be charged.

acute Denoting an angle that is less than a right angle; i.e. an angle less than $90°$ (or $\pi/2$ radian). *Compare* obtuse.

addend One of the numbers added together in a sum.

addition Symbol: + The operation of finding the sum of two or more quantities. In arithmetic, the addition of numbers is commutative $(4 + 5 = 5 + 4)$, associative $(2 + (3 + 4) = (2 + 3) + 4)$, and the identity element is zero $(5 + 0 = 5)$. The inverse operation to addition is subtraction. In *vector addition*, the direction of the two vectors affects the sum. Two vectors are added by placing them head-to-tail to form two sides of a triangle. The length and direction of the third side is the vector sum. *Matrix addition* can only be carried out between matrices with the same number of rows and columns, and the sum has the same dimensions. The elements in corresponding positions in each matrix are added arithmetically. *See also* matrix, sum, vector sum.

addition formulae Equations that express trigonometric functions of the sum or difference of two angles in terms of separate functions of the angles.
$$\sin(x + y) = \sin x \cos y + \cos x \sin y$$
$$\sin(x - y) = \sin x \cos y - \cos x \sin y$$
$$\cos(x + y) = \cos x \cos y - \sin x \sin y$$
$$\cos(x - y) = \cos x \cos y + \sin x \sin y$$
$$\tan(x + y) = (\tan x + \tan y)/(1 - \tan x \tan y)$$
$$\tan(x - y) = \tan x - \tan y)/(1 + \tan x \tan y)$$

They are used to simplify trigonometric expressions, for example, in solving an equation. From the addition formulae the following formulae can be derived:
The *double-angle formulae*:
$$\sin(2x) = 2 \sin x \cos x$$
$$\cos(2x) = \cos^2 x - \sin^2 x$$
$$\tan(2x) = 2\tan x/(1 - \tan^2 x)$$
The *half-angle formulae*:
$$\sin(x/2) = \pm\sqrt{[(1 - \cos x)/2]}$$
$$\cos(x/2) = \pm\sqrt{[(1 + \cos x)/2]}$$
$$\tan(x/2) = \sin x/(1 + \cos x) = (1 - \cos x)/\sin x$$
The *product formulae*:
$$\sin x \cos y = \tfrac{1}{2}[\sin(x + y) + \sin(x - y)]$$
$$\cos x \sin y = \tfrac{1}{2}[\sin(x + y) - \sin(x - y)]$$
$$\cos x \cos y = \tfrac{1}{2}[\cos(x + y) + \cos(x - y)]$$
$$\sin x \sin y = \tfrac{1}{2}[\cos(x - y) - \cos(x + y)]$$

addition of matrices *See* matrix.

address *See* store.

ad infinitum To infinity; an infinite number of times. Often abbreviated to *ad inf*.

adjacent 1. Denoting one of the sides forming a given angle in a triangle. In a right-angled triangle it is the side between the given angle and the right angle. In trigonometry, the ratios of this adjacent side to the other side lengths are used to define the cosine and tangent functions of the angle.
2. Denoting two sides of a polygon that share a common vertex.
3. Denoting two angles that share a common vertex and a common side.
4. Denoting two faces of a polyhedron that share a common edge.

adjoint (of a matrix) *See* cofactor.

admissible hypothesis Any hypothesis that could possibly be true; i.e. an hypothesis that has not been ruled out.

aether *See* ether.

affine geometry The study of properties left invariant under the group of affine transformations. *See* affine transformation, geometry.

affine transformation A transformation of the form

$x' = a_1x + b_1y + c_1, y' = a_2x + b_2y + c_2$

where $a_1b_2 - a_2b_1 = 0$. An affine transformation maps parallel lines into parallel lines, finite points into finite points, leaves the line at infinity fixed, and preserves the ratio of the distances separating three points that lie on a straight line. An affine transformation can always be factored into the product of the following important special cases:

(1) *translations*: $x' = x + a, y' = y + b$
(2) *rotations*: $x' = x\cos\theta + y\sin\theta, y' = -x\sin\theta + y\cos\theta$
(3) *stretchings* or *shrinkings*: $x' = tx, y' = ty$
(4) *reflections in the x-axis or y-axis*: $x' = x, y' = -y$ or $x' = -x, y' = y$
(5) *elongations* or *compressions*: $x' = x, y' = ty$ or $x' = tx, y' = y$

aleph The first letter of the Hebrew alphabet, used to denote transfinite cardinal numbers. $\aleph_0$, the smallest transfinite cardinal number, is the number of elements in the set of integers. $\aleph_1$ is the number of subsets of any set with $\aleph_0$ members. In general $\aleph_{n+1}$ is defined in the same way as the number of subsets of a set with $\aleph_n$ members.

algebra The branch of mathematics in which symbols are used to represent numbers or variables in arithmetical operations. For example, the relationship:

$3 \times (4 + 2) = (3 \times 4) + (3 \times 2)$

belongs to arithmetic. It applies only to this particular set of numbers. On the other hand the equation:

$x(y + z) = xy + xz$

is an expression in algebra. It is true for any three numbers denoted by x, y, and z. The above is a statement of the distributive law of arithmetic; similar statements can be written for the associative and commutative laws.

Much of elementary algebra consists of methods of manipulating equations to put them in a more convenient form. For example, the equation:

$x + 3y = 15$

can be changed by subtracting $3y$ from both sides of the equation, giving:

$x + 3y - 3y = 15 - 3y$
$x = 15 - 3y$

The effect is that of moving a term $(+3y)$ from one side of the equation to the other and changing the sign. Similarly a multiplication on one side of the equation becomes a division when the term is moved to the other side; for example:

$xy = 5$

becomes:

$x = 5/y$

'Ordinary' algebra is a generalization of arithmetic. Other forms of *higher algebra* also exist, concerned with mathematical entities other than numbers. For example, matrix algebra is concerned with the relations between matrices; vector algebra with vectors; Boolean algebra is applicable to logical propositions and to sets; etc. An algebra consists of a number of mathematical entities (e.g. matrices or sets) and operations (e.g. addition or set inclusion) with formal rules for the relationships between the mathematical entities. Such a system is called an *algebraic structure*.

algebra, Boolean *See* Boolean algebra.

ALGOL *See* program.

algorithm A mechanical procedure for performing a given calculation or solving a problem in a series of steps. One example is the common method of long division in steps. Another is the Euclidean algorithm for finding the highest common factor of two positive integers.

alternate angles A pair of equal angles formed by two parallel lines and a third line crossing both. For example, the two acute angles in the letter Z are alternate angles.

alternating series A series in which the terms are alternately positive and negative, for example:

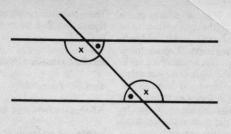

Alternate angles formed by a line
cutting two parallel lines.

$$S_n = -1 + 1/2 - 1/3 + 1/4 \ldots + (-1)^n/n$$

Such a series is convergent if the absolute value of each term is less than the preceding one. The example above is a convergent series.

An alternating series can be constructed from the sum of two series, one with positive terms and one with negative terms. In this case, if both are convergent separately then the alternating series is also convergent, even if the absolute value of each term is not always smaller than the one before it. For instance, the series:

$$S_1 = 1/2 + 1/4 + 1/8 + \ldots + 1/2^n$$

and

$$S_2 = -1/2 - 1/3 - 1/4 - 1/5 - \ldots (-1)/(n + 1)$$

are both convergent, and so their sum:

$$S_n = S_1 + S_2$$
$$= 1/2 - 1/2 + 1/4 - 1/3 + 1/8 - 1/4 + \ldots$$

is also convergent.

alternation *See* disjunction.

altitude The perpendicular distance from the base of a figure (e.g. a triangle, pyramid, or cone) to the vertex opposite.

ambiguous Having more than one possible meaning, value, or solution. For example, an ambiguous case occurs in finding the sides and angles of a triangle when two sides and an angle other then the included angle are known. One solution is an acute-angled triangle and the other is an obtuse-angled triangle.

ampere Symbol: A The SI base unit of electric current, defined as the constant current that, maintained in two straight parallel infinite conductors of negligible circular cross section placed one metre apart in vacuum, would produce a force between the conductors of 2×10^{-7} newton per metre.

amplitude The maximum value of a varying quantity from its mean or base value. In the case of a simple harmonic motion – a wave or vibration – it is half the maximum peak-to-peak value.

analog computer A type of computer in which numerical information (generally called data) is represented in the form of a quantity, usually a voltage, that can vary continuously. This varying quantity is an analogue of the actual data, i.e. it varies in the same manner as the data, but is easier to manipulate in the mathematical operations performed by the analog computer. The data is obtained from some process, experiment, etc.; it could be the changing temperature or pressure in a system or the varying speed of flow of a liquid. There may be several sets of data, each represented by a varying voltage.

The data is converted into its voltage analogue or analogues and calculations and other sorts of mathematical operations, especially the solution of differential equations, can then be performed on the voltage(s) (and hence on the data they represent). This is done by the user select-

..ng a group of electronic devices in the computer to which the voltage(s) are to be applied. These devices rapidly add voltages, and multiply them, integrate them, etc., as required. The resulting voltage is proportional to the result of the operation. It can be fed to a recording device to produce a graph or some other form of permanent record. Alternatively it can be used to control the process that produces the data entering the computer.

Analog computers operate in real time and are used, for example, in the automatic control of certain industrial processes and in a variety of scientific experiments. They can perform much more complicated mathematics than digital computers but are less accurate and are less flexible in the kind of things they can do. *See also* computer, hybrid computer.

analogy A general similarity between two problems or methods. Analogy is used to indicate the results of one problem from the known results of the other.

analysis The branch of mathematics concerned with the limit process and the concept of convergence. It includes the theory of differentiation, integration, infinite series, and analytic functions. Traditionally, it includes the study of functions of real and complex variables arising from differential and integral calculus.

analytic A function of a real or complex variable is *analytic* (or *holomorphic*) at a point if there is a neighbourhood N of this point such that the function is *differentiable* at every point of N. An alternative (and equivalent) definition is that a function is analytic at a point if it can be represented in a neighbourhood of this point by its Taylor series about it. A function is said to be analytic in a region if it is analytic at every point of that region.

analytical geometry (coordinate geometry) The use of coordinate systems and algebraic methods in geometry. In a plane Cartesian coordinate system a point is represented by a set of numbers and a curve is an equation for a set of points.

The geometric properties of curves and figures can thus be investigated by algebra.

anchor ring *See* torus.

and *See* conjunction.

AND gate *See* logic gate.

angle (plane angle) The spatial relationship between two straight lines. If two lines are parallel, the angle between them is zero. Angles are measured in degrees or, alternatively, in radians. A complete revolution is 360 degrees (360°). A straight line forms an angle of 180° and a right angle is 90°.

The angle between a line and a plane is the angle between the line and its orthogonal projection on the plane.

The angle between two planes is the angle between lines drawn perpendicular to the common edge from a point − one line in each plane. The angle between two intersecting curves is the angle between their tangents at the point of intersection.
See also solid angle.

ångstrom Symbol: Å A unit of length defined as 10^{-10} metre. The ångstrom is sometimes used for expressing wavelengths of light or ultraviolet radiation or for the sizes of molecules.

angular acceleration Symbol: α The rotational acceleration of an object about an axis; i.e. the rate of change of angular velocity with time:

$$\alpha = d\omega/dt$$

or

$$\alpha = d^2\theta/dt^2$$

where ω is angular velocity and θ is angular displacement. Angular acceleration is analogous to linear acceleration. *See* rotational motion.

angular displacement Symbol: θ The rotational displacement of an object about an axis. If the object (or a point on it) moves from point P_1 to point P_2 in a plane perpendicular to the axis, θ is the angle P_1OP_2, where O is the point at which

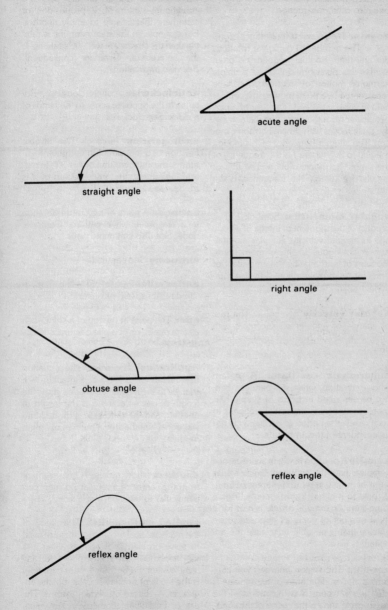

acute angle

straight angle

right angle

obtuse angle

reflex angle

reflex angle

Types of angles

the perpendicular plane meets the axis. *See also* rotational motion.

angular frequency (pulsatance) Symbol: ω The number of complete rotations per unit time. Angular frequency is often used to describe vibrations. Thus, a simple harmonic motion of frequency f can be represented by a point moving in a circular path at constant speed. The foot of a perpendicular from the point to a diameter of the circle moves with simple harmonic motion. The angular frequency of this motion is equal to $2\pi f$, where f is the frequency of the simple harmonic motion. The unit of angular frequency, like frequency, is the hertz.

angular momentum Symbol: L The product of the moment of inertia of a body and its angular velocity. Angular momentum is analogous to linear momentum, moment of inertia being the equivalent of mass for rotational motion. *See also* rotational motion.

angular velocity Symbol: ω The rate of change of angular displacement with time: $\omega = d\theta/dt$. *See also* rotational motion.

anharmonic oscillator A system whose vibration, while still periodic, cannot be described in terms of simple harmonic motions (i.e. sinusoidal motions). In such cases, the period of oscillation is not independent of the amplitude.

annuity A pension in which an insurance company pays the annuitant fixed regular sums of money in return for sums of money paid to it either in instalments or as a lump sum. An *annuity certain* is paid for a fixed number of years as opposed to an annuity that is payable only while the annuitant is alive.

annulus The region between two concentric circles. The area of an annulus is $\pi(R^2 - r^2)$, where R is the radius of the larger circle and r is the radius of the smaller.

antecedent In logic, the first part of a conditional statement; a proposition or statement that is said to imply another. For example, in the statement 'if it is raining then the streets are wet', 'it is raining' is the antecedent. *Compare* consequent. *See also* implication.

anticlockwise (counterclockwise) Rotating in the opposite sense to the hands of a clock. *See* clockwise.

antilogarithm (antilog) The inverse function of a logarithm. In common logarithms, the antilogarithm of x is 10^x. In natural logarithms, the antilogarithm of x is e^x. *See also* logarithm.

antinode A point of maximum vibration in a stationary wave pattern. *Compare* node. *See also* stationary wave.

antinomy *See* paradox.

antiparallel Parallel but acting in opposite directions, said of vectors.

apex The point at the top of a solid, such as a pyramid, or of a plane figure, such as a triangle.

Apollonius' theorem The equation that relates the length of a median in a triangle to the lengths of its sides. If a is the length of one side and b is the length of another, and the third side is divided into two equal lengths c by a median of length m, then:
$$a^2 + b^2 = 2m^2 + 2c^2$$

apothem (short radius) A line segment from the centre of a regular polygon perpendicular to the centre of a side.

applied mathematics The study of the mathematical techniques that are used to solve problems. Strictly speaking it is the application of mathematics to any 'real' system. For instance, pure geometry is the study of entities – lines, points, angles, etc. – based on certain axioms. The use of Euclidean geometry in surveying, architecture, navigation, or science is ap-

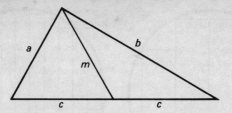

Appollonius' theorem: $a^2 + b^2 = 2m^2 + 2c^2$

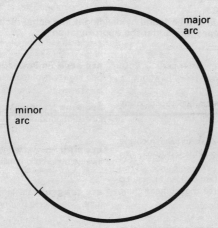

Arc: major and minor arcs of a circle

plied geometry. The term 'applied mathematics' is used especially for mechanics — the study of forces and motion. *Compare* pure mathematics.

approximate Describing a value of some quantity that is not exact but close enough to the correct value for some specific purpose, as within certain boundaries of error. It is also used as a verb meaning 'to find the value of a quantity within certain bounds of accuracy, but not exactly'. For example, one can approximate an ir-

rational number, such as π, by finding its decimal expansion to a certain number of places.

arc A part of a continuous curve. If the circumference of a circle is divided into two unequal parts, the smaller is known as the *minor arc* and the larger is known as the *major arc*.

arc cosecant (arc cosec, arc csc) An inverse cosecant. *See* inverse trigonometric functions.

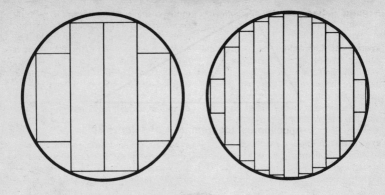

A curved area can be found by dividing it into equal rectangles. The more rectangles, the better the approximation.

arc cosech An inverse hyperbolic cosecant. *See* inverse hyperbolic functions.

arc cosh An inverse hyperbolic cosine. *See* inverse hyperbolic functions.

arc cosine (arc cos) An inverse cosine. *See* inverse trigonometric functions.

arc cotangent (arc cot) An inverse cotangent. *See* inverse trigonometric functions.

arc coth An inverse hyperbolic cotangent. *See* inverse hyperbolic functions.

Archimedes' principle The upward force of an object totally or partly submerged in a fluid is equal to the weight of fluid displaced by the object. The upward force, often called the *upthrust*, results from the fact that the pressure in a fluid (liquid or gas) increases with depth. If the object displaces a volume V of fluid of density ρ, then:

$$\text{upthrust } u = V\rho g$$

where g is the acceleration of free fall. If the upthrust on the object equals the object's weight, the object will float.

arc secant (arc sec) An inverse secant. *See* inverse trigonometric functions.

arc sech An inverse hyperbolic secant. *See* inverse hyperbolic functions.

arc sine (arc sin) An inverse sine. *See* inverse trigonometric functions.

arc sinh An inverse hyperbolic sine. *See* inverse hyperbolic functions.

arc tangent (arc tan) An inverse tangent. *See* inverse trigonometric functions.

arc tanh An inverse hyperbolic tangent. *See* inverse hyperbolic functions.

are A metric unit of area equal to 100 square metres. It is equivalent to 119.60 sq yd. *See also* hectare.

area Symbol: A The extent of a plane figure or a surface, measured in units of length squared. The SI unit of area is the square metre (m^2). The area of a rectangle is the product of its length and breadth. The area of a triangle is the product of the altitude and half the base. Closed figures bounded by straight lines have areas that can be determined by subdividing them into triangles. Areas for other figures can be found by using integral calculus.

Argand diagram *See* complex number.

argument (amplitude) In a complex number written in the form $r(\cos\theta + i\sin\theta)$, the angle θ is the argument. It is therefore the angle that the vector representing the complex number makes with the horizontal axis in an Argand diagram. *See also* complex number, modulus.

argument In logic, a sequence of propositions or statements, starting with a set of premisses (initial assumptions) and ending with a conclusion. *See also* logic.

arithmetic The study of the skills necessary to manipulate numbers in order to solve problems containing numerical information. It also involves an understanding of the structure of the number system and the facility to change numbers from one form to another; for example, the changing of fractions to decimals, and vice versa.

arithmetic and logic unit (ALU) *See* central processor.

arithmetic mean *See* mean.

arithmetic sequence (arithmetic progression) A sequence in which the difference between each term and the one after it is constant, for example, {9, 11, 13, 15, ... }. The difference between successive terms is called the *common difference*. The general formula for the nth term of an arithmetic sequence is:
$$n_n = a + (n - 1)d$$
where a is the first term of the sequence and d is the common difference. *Compare* geometric sequence. *See also* arithmetic series, sequence.

arithmetic series A series in which the difference between each term and the one after it is constant, for example, $3 + 7 + 11 + 15 + \ldots$. The general formula for an arithmetic series is:
$$S_n = a + (a + d) + (a + 2d) + \ldots$$
$$+ [a + (n - 1)d]$$
$$= \Sigma[a + (n - 1)d]$$

In the example, the first term, a, is 3, the *common difference*, d, is 4, and so the nth term, $a + (n - 1)d$, is $3 + (n - 1)4$. The sum to n terms of an arithmetic series is $n[2a + (n - 1)d]/2$ or $n(a + l)/2$ where l is the last (nth) term. *Compare* geometric series. *See also* series.

arm One of the lines forming a given angle.

array An ordered arrangement of numbers or other items of information, such as those in a list or table. In computing, an array has its own name, or *identifier*, and each member of the array is identified by a subscript used with the identifier. An array can be examined by a program and a particular item of information extracted by using this identifier and subscript.

artificial intelligence The branch of computer science concerned with programs that carry out tasks requiring intelligence when done by humans. Many of these tasks involve a lot more computation than is immediately apparent because much of the computation is unconscious in humans, making it hard to simulate. Programs now exist that play chess and other games at a high level, take decisions based on available evidence, prove theorems in certain branches of mathematics, recognize connected speech using a limited vocabulary, and use television cameras to recognize objects. Although these examples sound impressive, the programs have limited ability, no creativity, and each can only carry out a limited range of tasks. There is still a lot more research to be done before the ultimate goal of artificial intelligence is achieved, which is to understand intelligence well enough to make computers more intelligent than people.

assembler *See* program.

assembly language *See* program.

associative Denoting an operation that is independent of grouping. An operation · is associative if
$$a \cdot (b \cdot c) = (a \cdot b) \cdot c$$

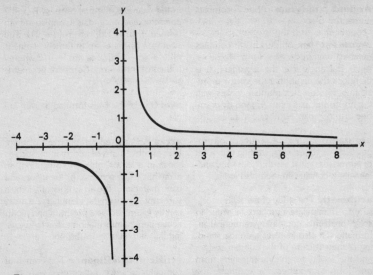

The x-axis and the y-axis are asymptotes to this curve.

for all values of *a*, *b*, and *c*. In ordinary arithmetic, addition and multiplication are associative operations. This is sometimes referred to as the *associative law of addition* and the *associative law of multiplication*. Subtraction and division are not associative. *See also* commutative, distributive.

astronomical unit (au, AU) The mean distance between the Sun and the Earth, used as a unit of distance in astronomy for measurements within the solar system. It is defined as 149 597 870 km.

asymmetrical Denoting any figure that cannot be divided into two parts that are mirror images of each other. The letter R, for example, is asymmetrical, as is any solid object that has a left-handed or right-handed characteristic. *Compare* symmetrical.

asymptote A straight line towards which a curve approaches but never meets. A hyperbola, for example, has two asymptotes. In two-dimensional Cartesian coordinates,

the curve with the equation $y = 1/x$ has the lines $x = 0$ and $y = 0$ as asymptotes, since *y* becomes infinitely small, but never reaches zero, as *x* increases, and vice versa.

atmosphere A unit of pressure equal to 760 mmHg. It is equivalent to 101 325 newtons per square metre (101 325 $N m^{-2}$).

atmospheric pressure *See* pressure of the atmosphere.

atto- Symbol: a A prefix denoting 10^{-18}. For example, 1 attometre (am) = 10^{-18} metre (m).

AU (au) *See* astronomical unit.

average *See* mean.

axial plane A fixed reference plane in a three-dimensional coordinate system. For example, in rectangular Cartesian coordinates, the planes defined by $x = 0$, $y = 0$, and $z = 0$ are axial planes. The x-coordinate of a point is its perpendicular distance from the plane $x = 0$, and the y and z coordinates are the perpendicular distances from the $y = 0$ and $z = 0$ planes respectively. *See also* coordinates.

axiom (postulate) In a mathematical or logical system, an initial proposition or statement that is accepted as true without proof and from which further statements, or theorems, can be derived. In a mathematical proof, the axioms are often well-known formulae for which the proof has already been established.

axiom of choice *See* choice, axiom of.

axis 1. A line about which a figure is symmetrical.
2. One of the fixed reference lines used in a graph or a coordinate system. *See* coordinates.
3. A line about which a curve or body rotates or revolves.
4. The line of intersection of two or more coaxial planes.

azimuth The angle θ measured in a horizontal plane from the x-axis in spherical polar coordinates. It is the same as the longitude of a point.

axial plane

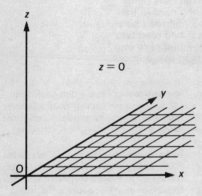

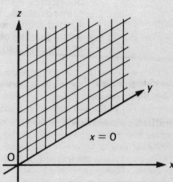

In three-dimensional rectangular Cartesian coordinates, the x and y axes lie in the axial plane $z = 0$, the y and z axes in the axial plane $x = 0$, and the x and z axes in the axial plane $y = 0$.

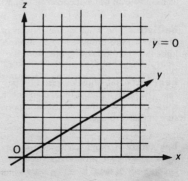

B

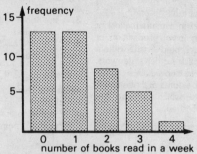

frequency

number of books read in a week

This bar chart shows the results
when a group of 40 school
students were asked how many
books they each had read in the
previous week. 13 had read none,
13 had read one, 8 had read two,
5 had read three, 1 had read four,
and no-one had read five or more.

backing store *See* store, magnetic
tape, disk.

ballistic pendulum A device for measuring the momentum (or velocity) of a
projectile (e.g. a bullet). It consists of a
heavy pendulum, which is struck by the
projectile. The momentum can be calculated by measuring the displacement of
the pendulum and using the law of constant momentum. If the mass of the projectile is known its velocity can be found.

ballistics The study of the motion of
objects that are propelled by an external
force (i.e. the motion of projectiles).

bar A unit of pressure defined as 10^5 pascals. The *millibar* (mb) is more common; it
is used for measuring atmospheric pressure in meteorology.

bar chart A graph consisting of bars
whose lengths are proportional to quantities in a set of data. It can be used when

one axis cannot have a numerical scale;
e.g. to show how many pink, red, yellow,
and white flowers grow from a packet of
mixed seeds. *See also* graph, histogram.

barn Symbol: b A unit of area defined as
10^{-28} square metre. The barn is sometimes used to express the effective cross-
sections of atoms or nuclei in the scatter-
ing or absorption of particles.

barrel A US unit of capacity used to measure solids. It is equal to 7056 cubic inches
(0.115 6 m^3).

barrel printer *See* line printer.

barycentre *See* centre of mass. Used in
particular for the centre of mass of a system of separate objects considered as a
whole.

barycentric coordinates Coordinates that relate the centre of mass of separate objects to the centre of mass of the
system of several objects as a whole. Con-

14

sider three objects with masses m_1, m_2, and m_3 with $m_1 + m_2 + m_3 = 1$, and their centres of mass at the points $p_1 = (x_1,y_1,z_1)$, $p_2 = (x_2,y_2,z_2)$, $p_3 = (x_3,y_3,z_3)$. Then the centre of mass of the three objects together is the point
$p = m_1p_1 + m_2p_2 + m_3p_3 = (m_1x_1 + m_2x_2 + m_3x_3, m_1y_1 + m_2y_2 + m_3y_3, m_1z_1 + m_2z_2 + m_3z_3)$
(m_1,m_2,m_3) are said to be the barycentric coordinates of the point p with respect to the points p_1, p_2, and p_3.

base 1. In geometry, the lower side of a triangle, or other plane figure, or the lower face of a pyramid or other solid. The altitude is measured from the base and at right angles to it.
2. In a number system, the number of different symbols used, including zero. For example, in the decimal number system, the base is ten. Ten units, ten tens, etc., are grouped together and represented by the figure 1 in the next position. In binary numbers, the base is two and the symbols used are 0 and 1.
3. In logarithms, the number that is raised to the power equal to the value of the logarithm. In common logarithms the base is 10; for example, the logarithm to the base 10 of 100 is 2:
$$\log_{10}100 = 2$$
$$100 = 10^2$$

base unit A unit that is defined in terms of reproducible physical phenomena or prototypes, rather than of other units. The second, for example, is a base unit in the SI, being defined in terms of the frequency of radiation associated with a particular atomic transition. Conventionally, seven units are chosen as base units in the SI. *See also* SI units.

BASIC *See* program.

basis vectors In two dimensions, two nonparallel vectors, scalar multiples of which are added to form any other vector in the same plane. For example, the unit vectors *i* and *j* in the directions of the x- and y-axes of a Cartesian coordinate sys-

tem are basis vectors. The position vector **OP** of the point P(2,3) is equal to $2\boldsymbol{i} + 3\boldsymbol{j}$. Similarly, in three dimensions a vector can be written as the sum of multiples of three basis vectors. *See* vector.

batch processing A method of operation, used especially in large computer systems, in which a number of programs are collected together and input to the computer as a single unit. The programs forming a batch can either be submitted at a central site or at a *remote job entry* site; there can be any number of remote job entry sites, which can be situated at considerable distance from the computer. The programs are then executed as time becomes available in the system. *Compare* time sharing.

Bayes' theorem A formula expressing the probability of an intersection of two or more sets as a product of the individual probabilities for each. It is used to calculate the probability that a particular event B_i has occurred when it is known that at least one of the set $\{B_1, B_2, \ldots B_n\}$ has occurred and that another event A has also occurred. This conditional probability is written as $P(B_i|A)$. $B_1, \ldots B_n$ form a partition of the sample space s such that $B_1 \cup B_2 \cup \ldots \cup B_n = s$ and $B_i \cap B_j = 0$, for all i and j. If the probabilities of $B_1, B_2, \ldots B_n$ and all of the conditional probabilities $P(A|B_j)$ are known, then $P(B_i|A)$ is given by
$$P(B_i)P(A|B_i)$$

beam compass *See* compasses.

bel *See* decibel.

Bernoulli trial An experiment in which there are two possible independent outcomes, for example, tossing a coin.

Bessel functions A set of functions, denoted by the letter J, that are solutions to Laplace's equation in cylindrical polar coordinates. The solutions form an infinite series and are listed in tables. *See also* partial differential equation.

15

beva- Symbol: B A prefix used in the USA to denote 10^9. It is equivalent to the SI prefix giga-.

bias A property of a statistical sample that makes it unrepresentative of the whole population. For example, if medical data is based on a survey of patients in a hospital, then the sample is a biased estimate of the general population, since healthy people will be excluded.

P Q	$P \equiv Q$
T T	T
T F	F
F T	F
F F	T

Biconditional

biconditional Symbol: $\leftrightarrow$ or $\equiv$ In logic, the relationship *if and only if* (often abbreviated to *iff*) that holds between a pair of propositions or statements P and Q only when they are both true or both false. It is also the relationship of logical *equivalence*; the truth of P is both a *necessary* and a *sufficient* condition for the truth of Q (and vice versa). The truth table definition for the biconditional is shown. *See* condition. *See also* truth table.

billion A number equal to 10^9 (i.e. one thousand million). This has always been the definition in the USA. In the UK a billion was formerly 10^{12} (i.e. one million million). In the 1960s, the British Treasury started using the term billion in the American sense of 10^9, and this usage has now supplanted the original 'million million'.

binary Denoting or based on two. A binary number is made up using only two different digits, 0 and 1, instead of ten in the decimal system. Each digit represents a unit, twos, fours, eights, sixteens, etc., instead of units, tens, hundreds, etc. For example, 2 is written as 10, 3 is 11, 16 (= 2^4) is 10 000. Computers calculate using binary numbers. The digits 1 and 0 correspond to on/off conditions in an electron-ic switching circuit or to presence or absence of magnetization on a disk or tape. *Compare* decimal, hexadecimal, octal, duodecimal.

binary logarithm A logarithm to the base two. The binary logarithm of 2 (written $\log_2 2$) is 1. *See* logarithm.

binary operation A mathematical procedure that combines two numbers, quantities, etc., to give a third. For example, multiplication of two numbers in arithmetic is a binary operation.

binomial An algebraic expression with two variables in it. For example, $2x + y$ and $4a + b = 0$ are binomials. *Compare* trinomial.

binomial coefficient The factor multiplying the variables in a term of a binomial expansion. For example, in $(x + y)^2 = x^2 + 2xy + y^2$ the binomial coefficients are 1, 2, and 1 respectively. *See* binomial expansion.

binomial distribution The distribution of the number of successes in an experiment in which there are two possible outcomes, i.e. success and failure. The probability of k successes is
$$b(k,n,p) = n!/k!(n-k)! \times p^n \times q^{n-k}$$
where p is the probability of success and q (= $1 - p$) the probability of failure on each trial. These probabilities are given by the terms in the binomial theorem expansion of $(p + q)^n$. The distribution has a mean np and variance npq. If n is large and p small it can be approximated by a Poisson distribution with mean np. If n is large and p is not near 0 or 1, it can be approximated by a normal distribution with mean np and variance npq.

binomial expansion A rule for the expansion of an expression of the form $(x + y)^n$. x and y can be any real numbers and n is an integer. The general formula, sometimes called the *binomial theorem*, is
$$(x + y)^n = x^n + nx^{n-1}y + [n(n-1)/2!]x^{n-2}y^2 + \ldots + y^n$$
This can be written in the form:

$$x^n + {}_nC_1x^{n-1}y + {}_nC_2x^{n-2}y^2 + \ldots$$
$$+ {}_nC_rx^{n-r}y^r + \ldots$$

The coefficients ${}_nC_1$, ${}_nC_2$, etc., are called *binomial coefficients*. In general, the rth coefficient, ${}_nC_r$, is given by

$$n!/(n - r)!r!$$

See also combination.

binomial theorem *See* binomial expansion.

birectangular Having two right angles. *See* spherical triangle.

bisector A straight line or a plane that divides a line, a plane, or an angle into two equal parts.

bistable circuit An electronic circuit that has two stable states. The circuit will remain in one state until the application of a suitable pulse, which will cause it to assume the other state.

Bistable circuits are used extensively in computer equipment for storing data and for counting. They usually have two input terminals to which pulses can be applied. A pulse on one input causes the circuit to change state; it will remain in that state until a pulse on the other input causes it to assume the alternative state. These circuits are often called *flip-flops*.

bit Abbreviation of *binary digit*, i.e. either of the digits 0 or 1 used in binary notation. Bits are the basic units of information in computing as they can represent the states of a two-valued system. For example, the passage of an electric pulse along a wire could be represented by a 1 while a 0 would mean that no pulse had passed. Again, the two states of magnetization of spots on a magnetic tape or disk can be represented by either a 1 or 0. *See also* binary, byte, word.

bivariate Containing two variable quantities. A plane vector, for example, is bivariate because it has both magnitude and direction.

A bivariate random variable (X, Y) has the joint probability $P(x,y)$; i.e. the probability that X and Y have the values x and y re-

spectively, is equal to $P(x) \times P(y)$, when X and Y are independent.

Board of Trade unit (BTU) A unit of energy equivalent to the kilowatt-hour (3.6 $\times 10^6$ joules). It was formerly used in the UK for the sale of electricity.

Bolzano–Weierstrass theorem The theorem that any bounded infinite set has an accumulation point. The accumulation point need not be in the set; e.g. the set $\{1, \frac{1}{2}, \frac{1}{4}, \ldots\}$ is bounded and infinite and it does have an accumulation point, namely 0, but that point is not in the set.

Boolean algebra A system of mathematical logic that uses symbols and set theory to represent logic operations in a mathematical form. It was the first system of logic using algebraic methods for combining symbols in proofs and deductions. Various systems have been developed and are used in probability theory and computing.

bound 1. In a set of numbers, a value beyond which there are no members of the set. A *lower bound* is less than or equal to every number in the set. An *upper bound* is greater than or equal to every number in the set. The *least upper bound* (or *supremum*) is the lowest of its upper bounds and the *greatest lower bound* (or *infimum*) is the largest of its lower bounds. For example, the set $\{0.6, 0.66, 0.666, \ldots\}$ has a least upper bound of $2/3$.
2. A bound of a function is a bound of the set of values that the function can take for the range of values of the variable. For example, if the variable x can be any real number, then the function $f(x) = x^2$ has a lower bound of 0.
3. In formal logic a variable is said to be bound if it is within the scope of a quantifier. For example, in the sentence $Fy \rightarrow (\exists x)Fx$, x is a bound variable, whereas y is not. A variable which is not bound is said to be *free*.

boundary condition In a differential equation, the value of the variables at a certain point, or information about their relationship at a point, that enables the arbitrary constants in the solution to be determined. For example, the equation

$$d^2y/dx^2 - 4dy/dx + 3y = 0$$

has a general solution

$$y = Ae^{-x} + Be^{-3x}$$

where A and B are arbitrary constants. If the boundary conditions are $y = 1$ at $x = 0$ and $dy/dx = 3$ at $x = 0$, the first can be substituted to obtain $B = 1 - A$. Differentiating the general solution for y gives

$$dy/dx = -Ae^{-x} - 3Be^{-3x}$$

and substituting the second boundary condition then gives

$$3 = -A - 3B = 2A - 3$$

That is, $A = 3$ and $B = -2$. *See also* differential equation.

branch 1. A section of a curve separated from the remainder of the curve by discontinuities or special points such as vertices, maximum points, minimum points, or cusps.
2. A departure from the normal sequential execution of instructions in a computer program. Control is thus transferred to another part of the program rather than passing in strict sequence from one instruction to the next. The branch will be either *unconditional*, i.e. it will always occur, or it will be *conditional*, i.e. the transfer of control will depend on the result of some arithmetical or logical test. *See also* loop.

breadth A horizontal distance, usually taken at right angles to a length.

Briggsian logarithm *See* logarithm.

British thermal unit (Btu) A unit of energy equal to $1.055\,06 \times 10^3$ joules. It was formerly defined by the heat needed

to raise the temperature of one pound of air-free water by one degree Fahrenheit at standard presure. Slightly different versions of the unit were in use depending on the temperatures between which the degree rise was measured.

BTU *See* Board of Trade unit.

Btu *See* British thermal unit.

buffer store (buffer) A small area of the main store of a computer in which information can be stored temporarily before, during, and after processing. A buffer can be used, for instance, between a peripheral device and the central processor, which operate at very different speeds. *See also* central processor, store.

bug An error or fault in a computer program. *See* debug.

buoyancy The tendency of an object to float. The term is sometimes also used for the upward force (upthrust) on a body. *See* centre of buoyancy. *See also* upthrust.

bushel A unit of capacity usually used for solid substances. In the UK it is equal to 8 UK gallons. In the USA it is equal to 64 US dry pints or 2150.42 cubic inches.

byte A subdivision of a word in computing, often being the number of bits used to represent a single letter, number, or other character. In most computers a byte consists of a fixed number of bits, usually eight. In some computers bytes can have their own individual addresses in the store. *See also* bit, character, word.

C

calculus (infinitesimal calculus) The branch of mathematics that deals with the differentiation and integration of functions. By treating continuous changes as if they consisted of infinitely small step changes, *differential calculus* can, for example, be used to find the rate at which the velocity of a body is changing with time (acceleration) at a particular instant.
Integral calculus is the reverse process, that is finding the end result of known continuous change. For example, if a car's acceleration a varies with time in a known way between times t_1 and t_2, then the total change in velocity is calculated by the integration of a over the time interval t_1 to t_2. *See* differentiation, integration.

calibration The marking of a scale on a measuring instrument. For example, a thermometer can be calibrated in degrees Celsius by marking the freezing point of water ($0°C$) and the boiling point of water ($100°C$).

calorie Symbol: cal A unit of energy approximately equal to 4.2 joules. It was formerly defined as the energy needed to raise the temperature of one gram of water by one degree Celsius. Because the specific thermal capacity of water changes with temperature, this definition is not precise. The mean or thermochemical calorie (cal_{TH}) is defined as 4.184 joules. The international table calorie (cal_{IT}) is defined as 4.186 8 joules. Formerly the mean calorie was defined as one hundredth of the heat needed to raise one gram of water from $0°C$ to $100°C$, and the $15°C$ calorie as the heat needed to raise it from $14.5°C$ to $15.5°C$.

cancellation Removing a common factor in a numerator and denominator, or removing the same quantity from both sides of an algebraic equation. For example, xy/yz can be simplified, by the cancellation of y, to x/z. The equation $z + x = 2$

$+ x$ is simplified to $z = 2$ by cancelling (subtracting) x from both sides.

candela Symbol: cd The SI base unit of luminous intensity, defined as the intensity (in the perpendicular direction) of the black-body radiation from a surface of $1/600\ 000$ square metre at the temperature of freezing platinum and at a pressure of $101\ 325$ pascals.

canonical form (normal form) In matrix algebra, the diagonal matrix derived by a series of transformations on another square matrix of the same order. *See also* diagonal matrix, square matrix.

Cantor's diagonal argument An argument to show that the real numbers, unlike the rationals, are not countable, and hence that there are more real numbers than rational numbers. The argument proceeds by assuming that the reals are countable and showing that this leads to a contradiction.
Every real number can be expressed as an infinite decimal expansion. We suppose that all reals are expressed in this way and since they are countable they can be arranged in order in a list as shown.
We now define a real number $b_1 . b_2 b_3 b_4 \ldots$ by saying that b_1 must be any number different from a_1. Hence our new number will not be equal to the first real number in our list. b_2 must be any number different from a_{21}, and hence the new number will not be equal to the second number on the list. Continuing in this way we have a method of producing an infinite decimal that must define a real number, but is not equal to any of the real numbers in our list. But we assumed that all real numbers occurred somewhere in the list; hence there is a contradiction, and so it cannot be true that the real numbers are countable.

Cantor set The set obtained by taking a closed interval in the real line, e.g. [0,1], and removing the middle third, i.e. the

$$\begin{pmatrix} 1 & 0 & 0 \\ 3 & 2 & 0 \\ 0 & 0 & 2 \end{pmatrix} \longrightarrow \begin{pmatrix} -3 & 0 & 0 \\ 3 & 2 & 0 \\ 0 & 0 & 2 \end{pmatrix} \begin{matrix} \text{multiply} \\ \text{row 1} \\ \text{by } -3 \end{matrix}$$

$$\begin{pmatrix} -3 & 0 & 0 \\ 3 & 2 & 0 \\ 0 & 0 & 2 \end{pmatrix} \longrightarrow \begin{pmatrix} -3 & 0 & 0 \\ 0 & 2 & 0 \\ 0 & 0 & 2 \end{pmatrix} \begin{matrix} \text{add row 1} \\ \text{to row 2} \end{matrix}$$

Reduction of a matrix to canonical form.

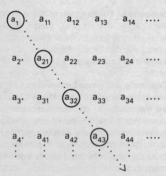

Cantor's diagonal argument

open interval $(1/3, 2/3)$, and then doing the same to the two remaining closed intervals $[0,1/3]$ and $[2/3,1]$, and so on *ad infinitum*. The set generated in this way has the remarkable property of containing uncountably many points – i.e. the same number of points as the whole real line – yet being nowhere dense – i.e. for any point in the set one can always find a point not in the set arbitrarily close to it. This set is also sometimes known as the *Cantor discontinuum*.

capital 1. The total sum of all the assets of a person or company, including cash, investments, household goods, land, buildings, machinery, and finished or unfinished goods.
2. A sum of money borrowed or lent on which interest is payable or received. *See* compound interest, simple interest.

3. The total amount of money contributed by the shareholders when a company is formed, or the amount contributed to a partnership by the partners.

carat (metric) A unit of mass used for precious stones. It is equal to 200 milligrams.

card A rectangular piece of stiff high-quality paper on which information can be recorded. In the case of a *punched card* the information is recorded as a pattern of rectangular holes in the paper. Each punched card is a uniform size and is divided into a number of columns along its length, usually 80. The 80-column card measures 18.73×8.25 cm. Each of the 80 columns has 12 positions at which a hole can be punched. A digit $(0-9)$, letter, or some other character is represented by a particular combination of holes in a column. Thus a number of adjacent columns can be used to record an item of information.
Punched cards were the earliest means by which information could be fed into and obtained from a computer. Their use is however decreasing. The information is usually recorded by means of a *keypunch*. This machine is manually operated from a keyboard similar to that of a typewriter, which causes the necessary holes to be punched in each card column. The accuracy of the punching is checked by a machine known as a *verifier*. The punched information is then fed into the computer using a *card reader*. This device senses the presence or absence of holes in each col-

umn and converts this information into a series of electric pulses. (A hole usually produces a pulse, a 'non-hole' produces no pulse.) The pulses are transmitted to the central processor of the computer. Although maybe 1000 cards can be read in a minute, the card reader is considered a very slow input device. Information is output on punched cards by means of a *card punch*, which automatically punches data into the cards. *Compare* paper tape, magnetic tape, disk, drum.

cardinal numbers Whole numbers that are used for counting or for specifying a total number of items, but not the order in which they are arranged. For example, when one says 'three books', the three is a cardinal number.

Two sets are said to have the same cardinal number if their elements can be put into one-to-one correspondence with each other. *Compare* ordinal numbers. *See also* aleph.

cardioid An epicycloid that has only one loop, formed by the path of a point on a circle rolling round the circumference of another that has the same radius. *See* epicycloid.

Cartesian coordinates A method of defining the position of a point by its distance from a fixed point (origin) in the direction of two or more straight lines. On a flat surface, two straight lines, called the *x*-axis and the *y*-axis, form the basis of a two-dimensional Cartesian coordinate system. The point at which they cross is the origin (O). An imaginary grid is formed by lines parallel to the axes and one unit length apart. The point (2,3), for example, is the point at which the line parallel to the *y*-axis two units in the direction of the *x*-axis, crosses the line parallel to the *x*-axis three units in the direction of the *y*-axis. Usually the *x*-axis is horizontal and the *y*-axis is perpendicular to it. These are known as *rectangular coordinates*. If the axes are not at right angles, they are *oblique coordinates*.

In three dimensions, a third axis, the *z*-axis, is added to define the height or depth of a point. The coordinates of a point are then three numbers (x,y,z). A right-handed system is one for which if the thumb of the right hand points along the *x*-axis, the fingers of the hand fold in the direction in which the *y*-axis would have to rotate to point in the same direction as the *z*-axis. A left-handed system is the mirror image of this. In a rectangular system, all three axes are mutually at right-angles. *See also* coordinates, polar coordinates.

Cartesian product The Cartesian product of two sets A and B, which is written $A \times B$, is the set of ordered pairs $<x,y>$ where x belongs to A and y belongs to B.
$$A \times B = \{<x,y> \mid x \in A \text{ and } y \in B\}$$

category A category consists of two classes: a class of objects and a class of *morphisms* – i.e. mappings that are in some sense structure-preserving. Associated with each pair of objects are a set of the morphisms and a law of composition for these morphisms. *Category theory* is the study of such entities. It provides a model for many situations where sets with certain structures are studied along with a class of mappings that preserve these structures. Examples of categories are sets with functions and groups with homomorphisms.

catenary The plane curve of a flexible uniform line suspended from two points. For example, an empty washing line attached to two poles and hanging freely between them follows a catenary. The catenary is symmetrical about an axis perpendicular to the line joining the two points of suspension. In Cartesian coordinates, the equation of a catenary that has its axis of symmetry lying along the *y*-axis at $y = a$, is
$$y = (a/2)(e^{x/a} + e^{-x/a})$$

catenoid The curved surface formed by rotating a catenary about its axis of symmetry.

Celsius degree Symbol: °C A unit of temperature difference equal to one hun-

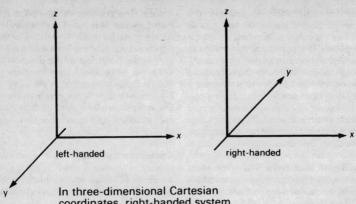

In three-dimensional Cartesian coordinates, right-handed system is the mirror image of a left-handed system.

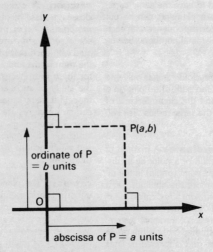

Two dimensional rectangular Cartesian coordinates showing a point P(a,b).

dredth of the difference between the temperatures of freezing and boiling water at one atmosphere pressure. It was formerly known as the degree centigrade and is equivalent to 1 K. On the Celsius scale water freezes at 0°C and boils at 100°C.

centi- Symbol: c A prefix denoting 10^{-2}. For example, 1 centimetre (cm) = 10^{-2} metre (m).

central conic A conic with a centre of symmetry; e.g. an ellipse or hyperbola.

central enlargement A central projection. *See also* scale factor.

central force A force that acts on any affected object along a line to an origin. For instance, the motion of electric forces between charged particles are central; frictional forces are not.

central processor (central processing unit, CPU) A highly complex electronic device that is the nerve centre of a computer. It consists of the *control unit* and the *arithmetic and logic unit* (ALU). Also sometimes considered part of the central processor is the main store, or memory, where a program or a section of a program is stored in binary form.
The control unit supervises all activities within the computer, interpreting the instructions that make up the program. Each instruction is automatically brought, in turn, from the main store and kept temporarily in a small store called a *register*. Electronic circuits analyse the instruction and determine the operation to be carried out and the exact location or locations in store of the data on which the operation is to be performed. The operation is actually performed by the ALU, again using electronic circuitry and a set of registers. It may be an arithmetical calculation, such as the addition of two numbers, or a logical operation, such as selecting or comparing data. This process of fetching, analysing, and executing instructions is repeated in the required order until an instruction to stop is executed.
The size of central processors has diminished considerably with advances in technology. It is now possible to form a central processor on a single silicon chip a few millimetres square in area, or on a small number of chips. This tiny device is known as a *microprocessor*. *See also* computer.

central projection (conical projection) A geometrical transformation in which a straight line from a point (called the *centre of projection*) to each point in the figure is continued to the point at which it passes through a second (image) plane. These points form the image of the

original figure. When a photographic image is created from a film using an enlarger, this is the kind of projection that takes place. The light source is at the centre of projection, the light rays are the straight lines, the film is the first plane, and the screen or point is the second. In this case the two planes are usually parallel, but this is not always so in central projection. *See also* projection.

centre A point about which a geometric figure is symmetrical.

centre of buoyancy For an object in a fluid, the centre of mass of the displaced volume of fluid. For a floating object to be stable the centre of mass of the object must lie below the centre of buoyancy; when the object is in equilibrium, the two lie on a vertical line. *See also* Archimedes' principle.

centre of curvature *See* curvature.

centre of gravity *See* centre of mass.

centre of mass (barycentre) A point in a body (or system) at which the whole mass of the body may be considered to act. Often the term *centre of gravity* is used. This is, strictly, not the same unless the body is in a constant gravitational field. The centre of gravity is the point at which the weight may be considered to act.
The centre of mass coincides with the centre of symmetry if the symmetrical body has a uniform density throughout. In other cases the principle of moments may be used to locate the point. For instance, two masses m_1 and m_2 a distance d apart have a centre of mass on the line between them. If this is a distance d_1 from m_1 and d_2 from m_2 then $m_1 d_1 = m_2 d_2$ or:
$$m_1 d_1 = m_2(d - d_1)$$
$$d_1 = m_2 d/(m_1 + m_2)$$
A more general relationship can be applied to a number of masses $m_1, m_2, \ldots, m_i$ that are respectively distances $r_1, r_2, \ldots, r_i$ from an origin. The distance r from the origin to the centre of mass is given by:
$$r = \Sigma r_i m_i / \Sigma m_i$$

centre of pressure

In the case of a body having a uniform density an integration must be used to obtain the position of the centre of mass, which coincides with the centroid. *See* centroid.

centre of pressure For a body or surface of a fluid, the point at which the resultant of pressure forces acts. If a surface lies horizontally in a fluid, the pressure at all points will be the same. The resultant force will then act through the centroid of the surface. If the surface is not horizontal, the pressure on it will vary with depth. The resultant force will now act through a different point and the centre of pressure is not at the centroid.

centre of projection The point at which all the lines forming a central projection meet. *See* central projection.

centrifugal force A force supposed to act radially outwards on a body moving in a curve. In fact there is no real force acting; centrifugal force is said to be a 'fictitious' force, and the use is best avoided. The idea arises from the effect of inertia on an object moving in a curve. If a car is moving around a bend, for instance, it is forced in a curved path by friction between the wheels and the road. Without this friction (which is directed towards the centre of the curve) the car would continue in a straight line. The driver also moves in the curve, constrained by friction with the seat, restraint from a seat belt, or a 'push' from the door. To the driver it appears that there is a force radially outwards pushing his body out –the centrifugal force. In fact this is not the case; if the driver fell out of the car he would move straight forward at a tangent to the curve. It is sometimes said that the centrifugal force is a 'reaction' to the centripetal force – this is not true. (The 'reaction' to the centripetal force is an outward push on the road surface by the tyres of the car.) *See also* centripetal force.

centripetal force A force that causes an object to move in a curved path rather than continuing in a straight line. The force is provided by, for instance:
– the tension of the string, for an object whirled on the end of a string;
– gravity, for an object in orbit round a planet;
– electric force, for an electron in the shell of an atom.
The centripetal force for an object of mass m with constant speed v, and path radius r is mv^2/r, or $m\omega^2 r$, where ω is angular speed. A body moving in a curved path has an acceleration because the direction of the velocity changes, even though the magnitude of the velocity may remain constant. This acceleration, which is directed towards the centre of the curve, is the *centripetal acceleration*. It is given by v^2/r or $\omega^2 r$.

centroid (mean centre) The point in a figure or solid at which the centre of mass would be if the figure or body were of uniform-density material. The centroid of a symmetrical figure is at the centre of symmetry; thus, the centroid of a circle is at its centre. The centroid of a triangle is the point at which the medians meet.
For non-symmetrical figures or bodies integration is used to find the centroid. The centroid of a line, figure, or solid is the point that has coordinates that are the mean values of the coordinates of all the points in the line, figure, etc. For a surface, the coordinates of the centroid are given by:
$$\bar{x} = [\iint x dx dy]/A, \text{ etc.}$$
the integration being over the surface, and A being the area. For a volume, a triple integral is used to obtain the coordinates of the centroid:
$$\bar{x} = [\iiint x dx dy dz]/V, \text{ etc.}$$
See also centre of mass.

c.g.s. system A system of units that uses the centimetre, the gram, and the second as the base mechanical units. Much early scientific work used this system, but it has now almost been abandoned.

chain A former unit of length equal to 22 yards. It is equivalent to 20.116 8 m.

chain rule A rule for expressing the derivative of a function $z = f(x)$ in terms of another function of the same variable, $u(x)$, where z is also a function of u. That is:
$$dz/dx = (dz/du)(du/dx)$$
This is often called the 'function of a function' rule.

For a function $z = f(x_1, x_2, x_3, \ldots)$ of several variables, in which each of the variables $x_1, x_2, x_3, \ldots$ is itself a function of a single variable, t, the derivative dz/dt, called the *total derivative*, is given by the chain rule for partial differentiation, which is:
$$dz/dt = (\partial z/\partial x_1)(dx_1/dt) +$$
$$(\partial z/\partial x_2)(dx_2/dt) + \ldots$$

channel A path along which information can travel in a computer system or communications system.

character One of a set of symbols that can be represented in a computer. It can be a letter, number, punctuation mark, or a special symbol. A character is stored or manipulated in the computer as a group of bits (i.e. binary digits). *See also* bit, byte, word, store.

characteristic 1. *See* logarithm.
2. *See* eliminant.

chi-square distribution (χ^2 distribution) The distribution of the sum of the squares of random variables with standard normal distributions. For example, if x_1, $x_2, \ldots x_i, \ldots$ are independent variables with standard normal distribution, then
$$\chi^2 = \Sigma x_i^2$$
has a chi-square distribution with n degrees of freedom, written χ_n^r. The mean and variance are n and $2n$ respectively. The values $\chi_n^2(\alpha)$ for which $P(\chi^2 \leqslant \chi_n^2(\alpha))$ $= \alpha$ are tabulated for various values of n.

chi-square test A measure of how well a theoretical probability distribution fits a set of data. For $i = 1, 2, \ldots m$ the value x_i occurs o_i times in the data and the theory predicts that it will occur e_i times. Provided $e_i \geqslant 5$ for all values of i (otherwise values must be combined), then
$$\chi^2 = \Sigma (o_i - e_i)^2/e_i$$

has a chi-square distribution with n degrees of freedom. *See also* chi-square distribution.

choice, axiom of The axiom states that, given any collection of sets, one can form a new set by *choosing* one element from each. This axiom may seem intuitively obvious and it was presupposed in many classical mathematical works. However, it has been a point of debate and controversy since many of its consequences appeared to be paradoxical. An example is the *Banach-Tarski theorem*, which proves that it is possible to cut a solid sphere into a finite number of pieces and to reassemble these pieces to form two solid spheres the same size as the original sphere. Despite these apparent paradoxes the axiom is widely accepted. It has many equivalents including the well-ordering principle and Zorn's lemma.

chord A straight line joining two points on a curve, for example, the line segment joining two points on the circumference of a circle.

circle The plane figure formed by a closed curve consisting of all the points that are at a fixed distance (the radius, r) from a particular point in the plane; the point is the centre of the circle. The diameter of a circle is twice its radius; the circumference is $2\pi r$; and its area is πr^2. In Cartesian coordinates, the equation of a circle centred at the origin is
$$x^2 + y^2 = r^2$$
The circle is the curve that encloses the largest possible area within any given perimeter length. It is a special case of an ellipse with eccentricity 0.

circular argument An argument that, tacitly or explicitly, assumes what it is trying to prove, and is consequently invalid.

circular cone A cone that has a circular base. *See* cone.

circular cylinder A cylinder in which the base is circular. *See* cylinder.

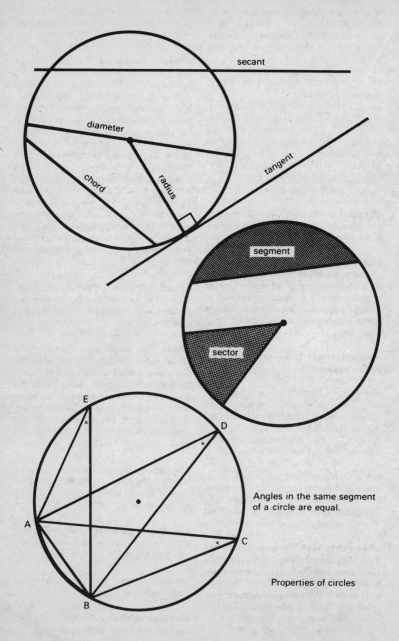

Angles in the same segment
of a circle are equal.

Properties of circles

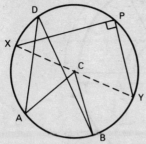

An angle that an arc subtends at the centre of a circle is twice the angle that it subtends at the circumference: $A\hat{C}B = 2A\hat{D}B$

An angle in a semicircle is a right angle: $X\hat{P}Y (= \frac{1}{2} X\hat{C}Y) = 90°$

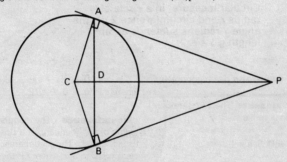

Two tangents from an external point:
(1) are equal, $PA = PB$
(2) subtend equal angles at the centre, $P\hat{C}A = P\hat{C}B$
(3) the line from the point to the centre bisects the line AB

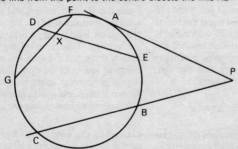

A tangent and a secant from an external point: $PC.PB = PA^2$

Two intersecting chords: $FX.GX = DX.XE$

Properties of circles

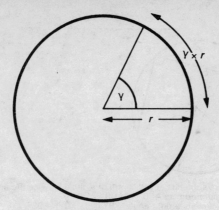

Circular measure: in a circle of
radius *r* and circumference 2 π*r*, the
angle ɣ radians subtends an arc
length g ɣ × *r*.

circular functions *See* trigonometry.

circular measure The measurement
of an angle in radians.

circular mil *See* mil.

circular motion A form of periodic (or
cyclic) motion; that of an object moving in
a circular path. For this to be possible, a
positive central force must act. If the object has a uniform speed *v* and the radius
of the circle is *r*, the angular velocity (ω) is
v/r. There is an acceleration towards the
centre of the circle (the centripetal acceleration) equal to v^2/r or $\omega^2 r$. *See also* centripetal force, rotational motion.

circumcentre *See* circumcircle.

circumcircle (circumscribed circle) The
circle that passes through all three vertices
of a triangle or through the vertices of any
other cyclic polygon. The figure inside the
circle is said to be *inscribed*. The point in
the figure that is the centre of the circle is
called the *circumcentre*. For a triangle with
side lengths *a*, *b*, and *c* the radius *r* of the
circumcircle is given by:

$$r = abc/\{4\sqrt{[s(s - a)(s - b)(s - c)]}\}$$
where *s* is $(a + b + c)/2$.

circumference The boundary, or
length of the boundary, of a closed curve,
usually a circle. The circumference of a circle is equal to $2\pi r$, where *r* is the radius of
the circle.

circumscribed Describing a geometric
figure that is drawn around and enclosing
another geometrical figure. For example,
in a square, a circle can be drawn through
the vertices. This is called the *circumscribed circle*, and the square is called the
inscribed square of the circle. Similarly, a
regular polyhedron might have a circumscribed sphere, and a rectangular pyramid
a circumscribed cone. *Compare* inscribed.

class 1. A grouping of data that is taken
as one item in a frequency table or histogram. *See also* frequency table, histogram.
2. Often used simply as a synonym for *set*.
However, in set theory it is sometimes de-
sirable, to avoid paradoxes, to allow the
existence of collections that are not sets.
Such collections are known as *classes* or

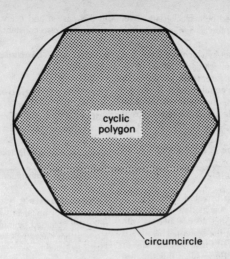

cyclic polygon

circumcircle

proper classes. For example, the collection of all sets is a proper class not a set.

classical mechanics A system of mechanics that is based on Newton's laws of motion. Relativity effects and quantum theory are not taken into account in classical mechanics.

class mark *See* frequency table.

clock pulse One of a series of regular pulses that are produced by an electronic device called a *clock* and are used to synchronize operations in a computer. Every instruction in a computer program causes a number of operations to be done by the central processor of the computer. Each of these operations, performed by the control unit or arithmetic and logic unit, are triggered by one clock pulse and must be completed before the next clock pulse. The interval at which the pulses occur is usually a few microseconds (millionths of a second). *See also* central processor.

clockwise Rotating in the same sense as the hands of a clock. For example, the head of an ordinary screw is turned clockwise (looking at the head of the screw) to drive in. Looking at the other end the rotation appears to be anticlockwise (counterclockwise).

closed Describing a set for which a given operation gives results in the same set. For example, the set of positive integers is closed with respect to addition and multiplication. Adding or multiplying any two members gives another positive integer. The set is not closed with respect to division since dividing certain integers does not give a positive integer (e.g. 4/5). The set of positive integers is also not closed with respect to subtraction (e.g. $5-7 = -2$). *See also* closed interval, closed set.

closed curve (closed contour) A curve, such as a circle or an ellipse, that forms a complete loop. It has no end points. A *simple* closed curve is a closed curve that does not cross itself. *Compare* open curve.

closed interval A set consisting of the numbers between two given numbers (end points), including the end points. For example, all the real numbers greater than or equal to 2 and less than or equal to 5 constitute a closed interval. The closed interval between two real numbers a and b is written $[a,b]$. On a number line the end points are marked by a blacked-in circle. *Compare* open interval. *See also* interval.

closed set A set in which the limits that define the set are included. The set of rational numbers greater than or equal to 0 and less than or equal to ten, written {x: $0 \leqslant x \leqslant 10; x \in R$}, and the set of points on and within a circle are examples of closed sets. *Compare* open sets.

closed surface A surface that has no boundary lines or curves, for example a sphere or an ellipsoid.

closed system (isolated system) A set of one or more objects that may interact with each other, but do not interact with the world outside the system. This means that there is no net force from outside or energy transfer. Because of this the system's angular momentum, energy, mass, and linear momentum remain constant.

closure *See* group.

cluster point *See* accumulation point.

coaxial 1. *Coaxial circles* are circles such that all pairs of the circles have the same radical axis. *See* radical axis.
2. *Coaxial planes* are planes that pass through the same straight line (the *axis*).

COBOL *See* program.

coding The writing of instructions in a computer programming language. The person doing the coding starts with a written description or a diagram representing the task to be carried out by the computer. This is then converted into a precise and ordered sequence of instructions in the language selected. *See also* flowchart, program.

coefficient A multiplying factor. For example, in the equation $2x^2 + 3x = 0$, where x is a variable, the coefficient of x^2 is 2 and the coefficient of x is 3. Sometimes the value of the coefficients is not known, although they are known to stay constant as x changes, for example, $ax^2 + bx = 0$.

In this case a and b are *constant coefficients*. *See also* constant.

coefficient, binomial *See* binomial coefficient.

coefficient of friction *See* friction.

cofactor The determinant of the matrix obtained by removing the row and column containing the element. The matrix formed by all the cofactors of the elements in a matrix is called the *adjoint* of the matrix. *See also* determinant.

coherent units A system or subset of units (e.g. SI units) in which the derived units are obtained by multiplying or dividing together base units, with no numerical factor involved.

colatitude *See* spherical polar coordinates.

collinear Lying on the same straight line. Any two points, for example, could be said to be collinear because there is a straight line that passes through both. Similarly, two vectors are collinear if they are parallel and both act through the same point.

cologarithm The logarithm of the reciprocal of a given number; i.e. the negative of the logarithm. It is sometimes used in logarithmic computation to avoid the use of negative mantissas or of subtraction of logarithms. *See* logarithm.

column matrix *See* column vector.

column vector (column matrix) A number (m) of quantities arranged in a single column; i.e. an $m \times 1$ matrix. For example, the vector that defines the displacement of the point (x,y,z) from the origin of a Cartesian coordinate system is usually written as a column vector.

combination Any subset of a given set of objects regardless of the order in which they are selected. If r objects are selected

$$A = \begin{pmatrix} a & b & c \\ d & e & f \\ g & h & i \end{pmatrix}$$

$$a' = \begin{vmatrix} e & f \\ h & i \end{vmatrix} = ei - hf$$

$$b' = \begin{vmatrix} d & f \\ g & i \end{vmatrix} = di - gf$$

$$c' = \begin{vmatrix} d & e \\ g & h \end{vmatrix} = dh - ge$$

The cofactors a', b', and c', of the elements a, b, and c in a 3 × 3 matrix A.

$$\begin{pmatrix} a' & b' & c' \\ d' & e' & f' \\ g' & h' & i' \end{pmatrix}$$

The adjoint of A.

The column vector that defines the displacement of a point (x, y, z) from the origin of a Cartesian coordinate system.

from n, and each object can only be chosen once, the number of different combinations is

$$n! / [r!(n-r)!]$$

written as $_nC_r$ or $C(n,r)$. For instance, if there are 15 students in a class and only 5 books, then each book has to be shared by 3 students. The number of ways in which this can happen – i.e. the number of combinations of 3 from 15 – is 15!/ 3!12!, or 455. If each object can be selected more than once the number of different combinations is $_{n+r-1}C_r$ *See also* factorial, permutation.

combinatorics (combinatorial analysis) The branch of mathematics that studies the number of possible configurations or arrangements of a certain type. It forms the basis for the theory of probability since we have to know how to calculate the total number of different ways an event *can* happen before we can hope to predict how it is *likely* to happen. There are many unsolved problems in combinatorics that at first appear simple. For example, a rectangular grid of some fixed dimension, *m* × *n*, is made up of unit squares each of one of two colours. How many different colour patterns are there if the number of boundary edges between the two colours is a certain fixed number? This problem was completely solved in two dimensions in the 1960s, but the solution for the three-dimensional problem is still unknown.

commensurable Able to be measured in the same way and in terms of the same units. For example, a 30 centimetre rule is commensurable with a 1 metre length of rope, because both can be measured in centimetres. Neither is commensurable with an area.

common denominator A whole number that is a common multiple of the denominators of two or more fractions. For example, 6 and 12 are both common denominators of 1/2 and 1/3. The *lowest* (or *least*) *common denominator* (LCD) is the smallest number that is a common multiple of the denominators of two or more fractions. For example, the LCD of 1/2, 1/3, and 1/4 is 12. Fractions are put in terms of the LCDs when they are to be added or subtracted:

$$1/2 + 1/3 + 1/4 = 6/12 + 4/12 + 3/12 = 13/12$$

common difference The difference between successive terms in an arithmetic sequence or arithmetic series.

common factor 1. A whole number that divides exactly into two or more given

numbers. For example, 7 is a common factor of 14, 49, and 84. Since 7 is the largest number that divides into all three exactly, it is the *highest common factor* (HCF). *See also* factor.
2. A number or variable by which several parts of an expression are multiplied. For example, in $4x^2 + 4y^2$, 4 is a common factor of x^2 and y^2, and from the distributive law for multiplication and addition,

$$4x^2 + 4y^2 = 4(x^2 + y^2)$$

common fraction *See* fraction.

common logarithm *See* logarithm.

common multiple A whole number that is a multiple of each of a group of numbers. For example, 100 is a common multiple of 5, 25, and 50. The *lowest* (or *least*) *common multiple* (LCM) is the smallest number that is a common multiple; in this case it is 50.

common ratio The ratio of successive terms in a geometric sequence or geometric series.

common tangent A single line that forms a tangent to two or more separate curves. The term is also used for the length of the line joining the two tangential points.

commutative Denoting an operation that is independent of the order of combination. A binary operation · is commutative if $a \cdot b = b \cdot a$ for all values of a and b. In ordinary arithmetic, multiplication and addition are commutative operations. This is sometimes referred to as the *commutative law of multiplication* and the *commutative law of addition*. Subtraction and division are not commutative operations. *See also* associative, distributive.

commutative group *See* Abelian group.

compact A set *S* of real numbers is compact if, given any collection of open sets whose union contains *S*, we can find a finite subcollection of those open sets

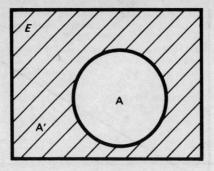

The shaded area in the Venn diagram is the complement A' of the set A.

whose union also contains S. The concept can be generalized to any topological space, and also to mathematical logic. In logic a formal system is said to be compact if it is such that, when a given sentence is a logical consequence of a given set of sentences, it is a consequence of some finite subset of them.

compasses An instrument used for drawing circles. It consists of two rigid arms joined by a hinge. At one end is a sharp point, which is placed at the centre of the circle. At the other end is a pencil or other marker, which traces out the circumference when the compasses are pivoted around the point. In a *beam compass*, used for drawing large circles, the sharp point and the marker are attached to opposite ends of a horizontal beam.

compiler See program.

complement The set of all the elements that are not in a particular set. If the set $A = \{1, 2, 3\}$, and the universal set, E, is taken as containing all the natural numbers, then the complement of A, written A' or $\bar{A}$, is $\{4, 5, 6, \ldots\}$. See Venn diagram.

complementary angles A pair of angles that add together to make a right an-gle (90° or $\pi/2$ radians). *Compare* conjugate angles, supplementary angles.

complete In mathematical logic a formal system is said to be complete if every true sentence in the system is also provable within the system. Not all logical systems have this property. *See* Gödel's incompleteness theorem.

completing the square A way of solving a quadratic equation, by dividing both sides by the coefficient of the square term and adding a constant, in order to express the equation as a single squared term. For example, to solve $3x^2 + 6x + 2 = 0$:
$$x^2 + 2x + \tfrac{2}{3} = 0$$
$$(x + 1)^2 - 1 + \tfrac{2}{3} = 0$$
$$x + 1 = +\sqrt{\tfrac{1}{3}} \text{ or } -\sqrt{\tfrac{1}{3}}$$
$$x = -1 + \sqrt{\tfrac{1}{3}} \text{ or } -1 - \sqrt{\tfrac{1}{3}}$$
See also quadratic equation.

complex fraction See fraction.

complex number A number that has both a real part and an imaginary part. The imaginary part is a multiple of the square root of minus one (i). Some algebraic equations cannot be solved with real numbers. For example, $x^2 + 4x + 6 = 0$ has the solutions $x = -2 + \sqrt{(-2)}$ and $x = -2 - \sqrt{(-2)}$. If the number system is extended to include $i = \sqrt{-1}$, all algebra-

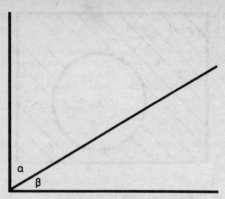

Complementary angles: α + β = 90°

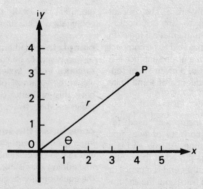

The point P(4,3) on an Argand
diagram represents the complex
number $z = 4 + 3i$. In the polar
form $z = r (\cos \Theta + i \sin \Theta)$.

ic equations can be solved. In this case the
solutions are $x = -2 + i\sqrt{2}$ and $x = -2 - i\sqrt{2}$. The real part is -2 and the imaginary part is $+i\sqrt{2}$ or $-i\sqrt{2}$.

Complex numbers are sometimes represented on an *Argand diagram*, which is similar to a graph in Cartesian coordinates, but with the horizontal axis representing the real part of the number and the vertical axis the imaginary part. Any complex number can also be written as a func-

tion of an angle θ, just as Cartesian coordinates can be converted into polar coordinates. Thus $r(\cos\theta + i \sin\theta)$ is equivalent to $x + iy$, where $x = r\cos\theta$ and $y = r\sin\theta$. Here, r is the *modulus* of the complex number and θ is the *argument* (or *amplitude*). This can also be written in the exponential form $r = e^i g$.

component forces *See* component
vectors.

34

component vectors The components of a given vector (such as a force or velocity) are two or more vectors with the same effect as the given vector. In other words the given vector is the resultant of the components. Any vector has an infinite number of sets of components. Some sets are more use than others in a given case, especially pairs at 90°. The component of a given vector (*V*) in a given direction is the projection of the vector on to that direction; i.e. $V\cos\theta$, where θ is the angle between the vector and the direction. *See* vector.

component velocities *See* component vectors.

composite number An integer that has more than one prime factor. For example, $4\,(=2\times2)$, $6\,(=2\times3)$, $10\,(=2\times5)$ are composite numbers. The prime numbers and ±1 are not composite.

composition The process of combining two or more functions to obtain a new one. For example the composition of $f(x)$ and $g(x)$, which is written $f\cdot g$, is obtained by applying $g(x)$ and then applying $f(x)$ to the result. If $f(x)=x-2$ and $g(x)=x^3+1$ then $f\cdot g(x)=f(x^3+1)=x^3-1$, whereas $g\cdot f(x)=g(x-2)=(x-2)^3+1$. As can be seen, these two resulting functions are not the same. In general, composing functions in a different order will produce different results.

compound interest The interest earned on capital, when the interest in each period is added to the original capital as it is earned. Thus the capital, and therefore the interest on it, increases year by year. If P is the principal (the original amount of money invested), R percent the interest rate per annum, and n the number of interest periods, then the compound interest is

$$P(1 + R/100)^n$$

This formula is a geometric progression whose first term is P (when $n=0$) and whose common ratio is $(1 + R/100)$. *Compare* simple interest.

compound proposition *See* proposition.

computability Intuitively, a problem or function is computable if it is capable of being solved by an ideal machine (computer) in a finite time. In the 1930s it was discovered that some problems had no algorithmic solution and could not be solved by computers. This led many mathematicians to try to formulate a precise definition of the intuitive concept of computability, and Turing, Gödel, and Church independently came up with three very different abstract definitions, which all turned out to define exactly the same set of functions. The definition given by Turing is that a function f is computable if for each element *x* of its domain, when some representation of *x* is placed on the tape of a Turing machine, the machine stops in a finite time with a representation of f(*x*) on the tape. *See* Turing machine.

computer Any automatic device or machine that can perform calculations and other operations on data. The data must be received in an acceptable form and is processed according to instructions. The most versatile and most widely used computer is the *digital computer*, which is usually referred to simply as a computer. *See also* analog computer, hybrid computer.

A digital computer is an automatically controlled calculating machine in which information, generally known as data, is represented by combinations of discrete electrical pulses denoted by the binary digits 0 and 1. Various operations, both arithmetical and logical, are performed on the data according to a set of instructions (a program). Instructions and data are fed into the main store or memory of the computer, where they are held until required. The instructions, coded like the data in binary form, are analysed and carried out by the central processor of the computer. The result of this processing is then delivered to the user.

The technology used in digital computers is so highly advanced that they can operate at extremely high speeds and can store a huge amount of information. The therm-

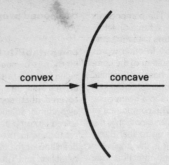

Concave and convex curvatures

ionic valves used in early computers were replaced by transistors; transistors, resistors, etc., were subsequently packed into integrated circuits, which have become more and more complicated. As the electronic circuits used in the various devices in a computer system have decreased in size and increased in complexity, so the computers themselves have grown smaller, faster, and more powerful. The *microcomputer* has been developed as a somewhat simpler version of the full-size *mainframe* computer. Computers now have an immense range of uses in science, technology, industry, commerce, education, and many other fields.

computer graphics In computing, the manipulation and presentation of computer information in pictorial as opposed to textual form. The information may range from a simple histogram to a complex annotated engineering design which may be rotated to allow the computer operator to view the design from all sides. The display is shown either on a visual display unit or on a printer or plotter. The terminal stores data as a set of discrete picture elements, or *pixels*. Each pixel may be individually addressed and the greater the number of pixels on a given screen the higher the resolution of the displayed image. The computer can be made to manipulate the information in many ways, such as straightening lines, moving or deleting areas of the display, and expanding details.

computer modelling The development of a description or mathematical representation (i.e. a *model*) of a complicated process or system, using a computer. This model can then be used to study the behaviour or control of the process or system by varying the conditions in it, again with the aid of a computer.

concave Curved inwards. For example, the inner surface of a hollow sphere is concave. Similarly in two dimensions, the inside edge of the circumference of a circle is concave. A *concave polygon* is a polygon that has one (or more) interior angles greater than 180°. *Compare* convex.

concentric Denoting circles or spheres that have the same centre. For example, a hollowed out sphere consists of two concentric spherical surfaces. *Compare* eccentric.

conclusion The proposition that is asserted at the end of an argument; i.e. what the argument sets out to prove.

condition In logic, a proposition or statement, P, that is required to be true in order that another proposition Q be true. If P is a *necessary condition* then Q could not be true without P. If P is a *sufficient condition*, then whenever P is true Q is also true, but not vice versa. For example, for a quadrilateral to be a rectangle it must satisfy the necessary condition that two of

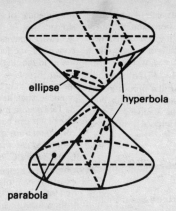

ellipse

hyperbola

parabola

The three sections of a cone – the ellipse, the parabola, and the hyperbola.

its sides be parallel, but this is not a sufficient condition. A sufficient condition for a quadrilateral to be a rhombus is that all its sides have a length of 5 centimetres, but this is not a necessary condition. For a rectangle to be a square it is both a necessary and a sufficient condition that all its sides are of equal length.

In formal terms, if P is a necessary condition for Q, then $Q \rightarrow P$. If P is a sufficient condition, then $P \rightarrow Q$. If P is a necessary and sufficient condition for Q then $P \equiv Q$. *See also* biconditional, symbolic logic.

conditional (conditional statement, conditional proposition) An *if* . . . *then* . . . statement.

conditional convergence *See* absolute convergence.

conditional equation *See* equation.

conditional probability *See* probability.

cone A solid defined by a closed plane curve (forming the base) and a point outside the plane (the vertex). A line segment from the vertex to a point on the plane curve generates a curved lateral sur-

face as the point moves around the plane curve. The line is the *generator* of the cone and the plane curve is its *directrix*. Any line segment from the vertex to the directrix is an *element* of the cone.

If the directrix is a circle the cone is a *circular cone*. If the base has a centre, a line from the vertex to this is an *axis* of the cone. If the axis is at right angles to the base the cone is a *right cone*; otherwise it is an *oblique cone*. The volume of a cone is one third of the base area multiplied by the altitude (the perpendicular distance from the vertex to the base). For a right circular cone

$$V = \pi r^2 h / 3$$

where r is the radius of the base and h the altitude. The area of the curved (lateral) surface of a right circular cone is $\pi r s$, where s is the length of an element (the *slant height*).

If an extended line is used to generate the curved surface (i.e. extending beyond the directrix and beyond the vertex), an extended surface is produced with two parts (*nappes*) each side of the vertex. More strictly, this is called a *conical surface*.

confidence interval An interval that is thought, with a preselected degree of confidence, to contain the value of a parameter being estimated. For example, in a binomial experiment the $\alpha\%$ confidence interval for the probability of success P lies between $P - a$ and $P + a$, where

$$a = z\sqrt{[P(1 - P)/N]}$$

N is the sample size, P the proportion of successes in the sample, and z is given by a table of area under the standard normal curve. P will lie in this interval α times out of every 100.

confocal conics Two or more conics that have the same focus.

conformable matrices *See* matrix.

conformal mapping A geometrical transformation that does not change the angles of intersection between two lines or curves. For example, Mercator's projection is a conformal mapping in which any angle between a line on the spherical surface and a line of latitude or longitude will be the same on the map.

congruence The property of being congruent.

congruent 1. Denoting two or more figures that are identical in size and shape. Two congruent plane figures will fit into the area occupied by each other; i.e. one could be brought into coincidence with the other by moving it without change of size. Two circles are congruent if they have the same radius. The conditions for two triangles to be congruent are:
(1) Two sides and the included angle of one are equal to two sides and the included angle of the other.
(2) Two angles and the included side of one are equal to two angles and the included side of the other.
(3) Three sides of one are equal to three sides of the other.
In solid geometry, two figures are congruent if they can be brought into coincidence in space.
Sometimes the term *directly congruent* is used to describe identical figures; *indirectly congruent* figures are ones that are mirror images of each other. *Compare* similar.

2. Two elements a and b of a ring are *congruent modulo* d if there exist elements in the ring, p, q, and r, such that $a = dp + r$, $b = dq + r$. Intuitively, this means that they both leave the same remainder when divided by d.

3. Two square matrices A and B are congruent if A can be transformed into B by a *congruent transformation*; i.e. there exists a nonsingular matrix C such that $B = C^{T}AC$, where C^{T} is the transpose of C.

conic A type of plane curve defined so that for all points on the curve the distance from a fixed point (the *focus*) has a constant ratio to the perpendicular distance from a fixed straight line (the *directrix*). The ratio is the eccentricity of the conic, e; i.e. the eccentricity is the distance from curve to focus divided by distance from curve to directrix.
The type of conic depends on the value of e: when e is less than 1 it is an ellipse; when e equals 1 it is parabola; when e is greater than 1 it is hyperbola. A circle is a special case of an ellipse with eccentricity 0.
The original definition of conics was as plane sections of a conical surface – hence the name *conic section*. In a conical surface having an apex angle of 2θ, the cross-section on a plane that makes an angle θ with the axis of the cone, (i.e. a plane parallel to the slanting edge of the cone) is a parabola. A cross-section in a plane that makes an angle greater than θ with the axis is an ellipse. A cross-section in a plane making an angle less than θ with the axis is a hyperbola and because this plane cuts both halves (nappes) of the cone, the hyperbola has two arms.
There are various ways of writing the equation of a conic. In Cartesian coordinates:

$$(1 - e^2)x^2 + 2e^2qx + y^2 = e^2q$$

where the focus is at the origin and the directrix is the line $x = q$ (a line parallel to the y-axis a distance q from the origin). The general equation of a conic (i.e. the *general conic*) is:

$$ax^2 + bxy + cy^2 + dx + ey + f = 0$$

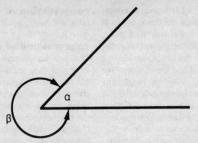

Conjugate angles: $\alpha + \beta = 360°$

where a, b, c, d, e, and f are constants (here e is *not* the eccentricity). This includes degenerate cases (*degenerate conics*), such as a point, a straight line, and a pair of intersecting straight lines. A point, for example, is a section through the vertex of the conical surface. A pair of intersecting straight lines is a section down the axis of the surface. The tangent to the general conic at the point (x_1, y_1) is:

$$ax_1x + b(xy_1 + x_1y) + cy_1y + d(x + x_1) + e(y + y_1) + f = 0$$

See also ellipse, hyperbola, parabola.

conical helix See helix.

conical projection See central projection.

conical surface See cone.

conic sections See conic.

conjugate angles A pair of angles that add together to make a complete revolution (360° or 2π radians). *Compare* complementary angles, supplementary angles.

conjugate axis See hyperbola.

conjugate complex numbers Two complex numbers of the form $x + iy$ and $x - iy$, which when multiplied together have a real product $x^2 + y^2$. If $z = x + iy$, the complex conjugate of z is $\bar{z} = x - iy$.

conjugate hyperbola See hyperbola.

P	Q	$P \wedge Q$
T	T	T
T	F	F
F	T	F
F	F	F

conjunction

conjunction Symbol: $\wedge$ In logic, the relationship *and* between two or more propositions or statements. The conjunction of P and Q is true when P is true and Q is true, and false otherwise. The truth table definition for conjunction is shown in the illustration. *Compare* disjunction. *See also* truth table.

connected Intuitively, a connected set is a set with only one piece. More rigorously, a set in a topological space is said to be connected if it is not the union of two nonempty disjoint closed sets. For example, the set of rational numbers is not connected since the set of all rational numbers less than $\sqrt{3}$ and the set of all rational numbers greater than $\sqrt{3}$ are both closed in the set of all rational numbers. However, the set of all real numbers is connected since no such decomposition is possible. A subset S of a topological space is *path-connected* if any two points in S can be joined by a *path* in S, where a path from a to b in S is a continuous map $f:[0,1] \rightarrow S$ such that $f(0) = a$ and $f(1) = b$.

A *simply connected* set is a path-connected set such that any closed curve within it can be deformed continuously to a point of the set without leaving the set. A path-connected set that is not simply connected is *multiply connected* and the *connectivity* of the set is one plus the maximum number of points that can be deleted if the set is a curve, or one plus the maximum number of closed cuts that can be made if the set is a surface, without separating the set so that it is no longer path-connected. For example, the region between two concentric circles has *connectivity two*, or is *doubly connected*, since one closed cut can be made which still leaves a connected region.

connectivity The number of cuts needed to break a shape in two parts. For example, a rectangle, a circle, and a sphere, all have a connectivity of one. A flat disc with a hole in it or a torus has a connectivity of two. *See also* topology.

consequent In logic, the second part of a conditional statement; a proposition or statement that is said to follow from or be implied by another. For example, in the statement 'if Jill is happy, then Jack is happy', 'Jack is happy' is the consequent. *Compare* antecedent. *See also* implication.

conservation law A law stating that the total value of some physical quantity is conserved (i.e. remains constant) throughout any changes in a closed system. The conservation laws applying in mechanics are the laws of constant mass, constant energy, constant linear momentum, and constant angular momentum.

conservation of angular momentum, law of *See* constant angular momentum, law of.

conservation of energy, law of *See* constant energy, law of.

conservation of (linear) momentum, law of *See* constant linear momentum, law of.

conservation of mass, law of *See* constant mass, law of.

conservation of mass and energy The law that the total energy (rest mass energy + kinetic energy + potential energy) of a closed system is constant. In most chemical and physical interactions the mass change is undetectably small, so that the measurable rest-mass energy does not change (it is regarded as 'passive'). The law then becomes the classical *law of conservation of energy*. In practice, the inclusion of mass in the calculation is necessary only in the case of nuclear changes or systems involving very high speeds. *See also* mass-energy equation, rest mass.

conservation of momentum, law of *See* constant linear momentum, law of.

conservative field A field such that the work done in moving an object between two points in the field is independent of the path taken. *See* conservative force.

conservative force A force such that, if it moves an object between two points, the energy transfer (work done) does not depend on the path between the points. It must then be true that if a conservative force moves an object in a closed path (back to the starting point), the energy transfer is zero. Gravitation is an example of a conservative force; friction is a non-conservative force.

consistent Describing a theory, system, or set of propositions giving rise to no contradictions. Arithmetic, for example, is thought to be a consistent logical system because none of its axioms nor any of the theorems that are derived from these by the rules, are believed to be contradictory. *See* contradiction.

consistent equations A set of equations that can be satisfied by at least one set of values for the variables. For example, the equations $x + y = 2$ and $x + 4y = 6$ are satisfied by $x = 2/3$ and $y = 4/3$ and they are therefore consistent. The equations $x + y = 4$ and $x + y = 9$ are inconsistent.

constant A quantity that does not change its value in a general relationship between variables. For example, in the equation $y = 2x + 3$, where x and y are variables, the numbers 2 and 3 are constants. In this case they are *absolute constants* because their values never change. Sometimes a constant can take any one of a number of values in different applications of a general formula. In the general quadratic equation
$$ax^2 + bx + c = 0$$
a, b, and c are *arbitrary constants* because no values are specified for them. An indefinite integral includes an arbitrary constant (the *constant of integration*), which depends on the limits chosen. *See also* indefinite integral.

constant angular momentum, law of (law of conservation of angular momentum) The principle that the total angular momentum of a system cannot change unless a net outside torque acts on the system. *See also* constant linear momentum, law of.

constant energy, law of (law of conservation of energy) The principle that the total energy of a system cannot change unless energy is taken from or given to the outside. *See also* mass-energy equation.

constant linear momentum, law of (law of conservation of (linear) momentum) The principle that the total linear momentum of a system cannot change unless a net outside force acts.

constant mass, law of (law of conservation of mass) The principle that the total mass of a system cannot change unless mass is taken from or given to the outside. *See also* mass-energy equation.

constant momentum, law of *See* constant linear momentum, law of.

construct In geometry, to draw a figure, line, point, etc., meeting certain conditions; e.g. a line that bisects a given line. Usually certain specific restrictions are imposed on the method used; e.g. using only a straight edge and compasses. There is an important class of problems concerning questions of whether certain things can be constructed using given methods. Examples are two celebrated problems of whether it is possible to construct two lines that trisect a given angle, and to construct a square equal in area to a given circle – in both cases using only a straight edge and compasses. Both these constructions have been shown to be impossible.

constructive proof A proof that not only shows that a certain mathematical entity, such as a root of an equation or a fixed-point of a transformation, exists, but also explicitly produces it. Constructive proofs are usually considerably longer and more complicated and harder to find than nonconstructive proofs of the same results. Many results that have been proved nonconstructively have yet to be given constructive proofs.

constructivist mathematics An approach to mathematics that insists that only constructive proofs are acceptable and rejects as meaningless nonconstructive proofs. Constructivist mathematics is considerably more restricted than classical mathematics and rejects many of its theorems. Different varieties of constructivism differ over what exactly counts as an acceptably constructive proof. One of the best known examples of mathematical constructivism is the intuitionism of Brouwer.

continued product Symbol: Π The product of a number of related terms. For example, $2 \times 4 \times 6 \times 8 \ldots$ is a continued product, written:
$$\prod^{k} a_n$$
This means the product of k terms, with the nth term, $a_n = 2n$.
$$\prod_{1}^{\infty} a_n$$
has an infinite number of terms.

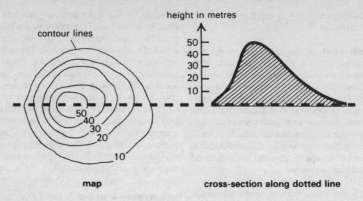

A hill shown as contour lines on a map and as a cross-section.

continuous function A function that has no sudden changes in values as the variable increases or decreases smoothly. More precisely, a function f(x) is continuous at a point $x = a$ if the limit of f(x) as x approaches a is f(a). When a function does not satisfy this condition at a point, it is said to be *discontinuous*, or to have a *discontinuity*, at that point. For example, tanθ has discontinuities at $\theta = \pi/2, 3\pi/2, 5\pi/2, \ldots$. A function is continuous in an interval of x if there are no points of discontinuity in that interval.

continuum A compact connected set with at least two points. The conditions that the set has at least two points and is connected imply that the set has an infinite number of points. Any closed interval of the real numbers is a continuum and the set of all real numbers is called the *real continuum*.

The *continuum hypothesis* is the conjecture that every infinite subset of the real continuum has the cardinal number either of the positive integers or of the entire set of real numbers. This is equivalent to the statement that $2^{\aleph_0}$ is the least cardinal number greater than $\aleph$. *See* aleph, cardinal number.

contour integral *See* line integral.

contour line A line on a map joining points of equal height. Contour lines are usually drawn for equal intervals of height, so that the steeper a slope, the closer together the contour lines.

contradiction In logic, a proposition, statement, or sentence that both asserts something *and* denies it. It is a form of words or symbols that cannot possibly be true; for example, 'if I can read the book then I cannot read the book' and 'he is coming and he is not coming'. *Compare* tautology. *See also* logic.

contradiction, law of *See* laws of thought.

contrapositive In logic, a statement in which the antecedent and consequent of a conditional are reversed and negated. The contrapositive of $A \rightarrow B$ is $\sim B \rightarrow \sim A$ (not B implies not A), and the two statements are logically equivalent. *See* implication. *See also* biconditional.

control unit *See* central processor.

convergent sequence A sequence in which the difference between each term and the one following it becomes smaller throughout the sequence; i.e. the difference between the nth term and the (n +

42

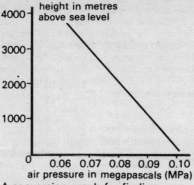

A conversion graph for finding altitude from air pressure measurements. (Standard air pressure at sea level is 1.01325 million pascals.)

1)th term decreases as n increases. For example, $\{1, \frac{1}{2}, \frac{1}{4}, \frac{1}{8}, \ldots\}$ is a convergent sequence, but $\{1, 2, 4, 8, \ldots\}$ is not. A convergent sequence has a limit; i.e. a value towards which the nth term tends as n becomes infinitely large. In the first example here the limit is 0. *Compare* divergent sequence. *See also* convergent series, geometric sequence, sequence.

convergent series An infinite series $a_1 + a_2 + \ldots$ is *convergent* if the partial sums $a_1 + a_2 + \ldots + a_n$ tend to a limit value as n tends to infinity. For example, the series $S = 1 + \frac{1}{2} + \frac{1}{4} + \frac{1}{8} + \ldots$ is a convergent series with sum 2, since 2 is the limit approached by the sum of the first n terms, namely $1 - (1/2^n)$ as n tends to infinity. The series $1 + (-1) + 1 + (-1) + 1 + \ldots$ is not convergent. *Compare* divergent series. *See also* convergent sequence, geometric series, series.

converse A logical implication taken in the reverse order. For example, the converse of
if I am under 16, then I go to school
is
if I go to school, then I am under 16.
The converse of an implication is not always true if the implication itself is true.

There are a number of theorems in mathematics for which both the statement and the converse are true. For example, the theorem:
if two chords of a circle are equidistant from the centre, then they are equal
has a true converse:
if two chords of a circle are equal, then they are equidistant from the centre of the circle.
See also implication.

conversion factor The ratio of a measurement in one set of units to the equivalent numerical value in other units. For example, the conversion factor from inches to centimetres is 2.54 because 1 inch = 2.54 centimetres (to two decimal places).

conversion graph A graph showing a relationship between two variable quantities. If one quantity is known, the corresponding value of the other can be read directly from the graph. For example, air pressure depends on height above sea level. A standard curve of altitude against air pressure may be plotted on a graph. An air-pressure measurement can then be converted to an indication of height by reading the appropriate value from the graph.

convex Curved outwards. For example, the outer surface of a sphere is convex. Similarly, in two dimensions, the outside of a circle is its convex side. A *convex polygon* is one in which no interior angle is greater than 180°. *Compare* concave.

coordinate geometry *See* analytical geometry.

coordinates Numbers that define the position of a point, or set of points. A fixed point, called the *origin*, and fixed lines, called *axes*, are used as a reference. For example, a horizontal line and a vertical line drawn on a page might be defined as the x-axis and the y-axis respectively, and the point at which they cross as the origin (O). Any point on the page can then be given two numbers – its distance from O along the x-axis from left to right and its distance upwards from O in the direction of the y-axis. These two numbers would be the x and y coordinates of the point. This type of coordinate system is known as a rectangular Cartesian coordinate system. It can have two axes, as on a flat surface, such as a map, or three axes, when depth or height also have to be specified. Another type of coordinate system (polar coordinates) expresses the position of a point as radial distance from the origin (the *pole*), with its direction expressed as an angle or angles (positive when anticlockwise) between the radius and a fixed axis (the *polar axis*). *See also* Cartesian coordinates, polar coordinates.

coplanar Lying in the same plane. Any set of three points, for example, could be said to be coplanar because there is a plane in which they all lie. Two vectors are coplanar if there is a plane that contains both.

coplanar forces Forces in a single plane. If only two forces act through a point, they must be coplanar. So too are two parallel forces. However, nonparallel forces that do not act through a point cannot be coplanar. Three or more nonparallel forces acting through a point may not be coplanar. If a set of coplanar forces act

on a body, their algebraic sum must be zero (i.e. the resultant in one direction must equal the resultant in the opposite direction). In addition there must be no couple on the body (the moment of the forces about a point must be zero).

Coriolis force A 'fictitious' force used to describe the motion of an object in a rotating system. For instance, air moving from north to south over the surface of the Earth would, to an observer outside the Earth, be moving in a straight line. To an observer on the Earth the path would appear to be curved, as the Earth rotates. Such systems can be described by introducing a tangential Coriolis 'force'. The idea is used in meteorology to explain wind directions.

corollary A result which follows easily from a given theorem, so that it is not necessary to prove it as a separate theorem.

correction A quantity added to a previously obtained approximation to yield a better approximation. When using logarithmic or trigonometric tables the correction is the number added to a logarithm or to a trigonometric function in the table to give the logarithm or trigonometric function of a number or angle that is not in the table.

correlation In statistics, the *correlation coefficient* of two random variables X and Y is defined by
$$r(X,Y) = \text{cov}(X,Y)/\sqrt{(\text{var}(X)\text{var}(Y))}$$
where cov and var denote covariance and variance respectively. It satisfies $-1 \leqslant r \leqslant 1$ and is a measurement of the interdependence between random variables, or their tendency to vary together. If $r \neq 0$ then X and Y are said to be *correlated*: they are correlated *positively* if $0 < r \leqslant 1$ and *negatively* if $-1 \leqslant r < 0$. If $r = 0$ then X and Y are said to be *uncorrelated*.

For two sets of numbers $(x_1, \ldots x_n)$ and $(y_1, \ldots y_n)$ the correlation coefficient is
$$r = \frac{\sum_{i=1}^{n}(x_i - \bar{x})(y_i - \bar{y})}{\sqrt{\left[\sum_{i=1}^{n}(x_i - \bar{x})^2 \sum_{i=1}^{n}(y_i - \bar{y})^2\right]}}$$

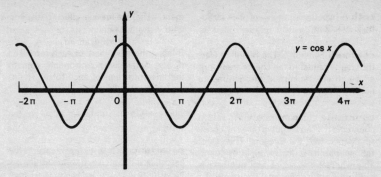

The graph of $y = \cos x$ with x in radians

where $\bar{x}$ and $\bar{y}$ are the corresponding means. It measures how near the points $(x_1, y_1) \ldots (x_n, y_n)$ are to lying on a straight line. If $r = 1$ the points lie on a line and the two sets of data are said to be in *perfect correlation*.

correspondence See function.

cos See cosine.

cosec See cosecant.

cosecant (cosec, csc) A trigonometric function of an angle equal to the reciprocal of its sine; i.e. $\text{cosec}\,\alpha = 1/\sin\alpha$. *See also* trigonometry.

cosech A hyperbolic cosecant. *See* hyperbolic functions.

cosh A hyperbolic cosine. *See* hyperbolic functions.

cosine (cos) A trigonometric function of an angle. The cosine of an angle α (cosα) in a right-angled triangle is the ratio of the side adjacent to it, to the hypotenuse. This definition applies only to angles between $0°$ and $90°$ (0 and $\pi/2$ radians). More generally, in rectangular Cartesian coordinates, the x-coordinate of any point on the circumference of a circle of radius r cen-

tred on the origin is $r\cos\alpha$, where α is the angle between the x-axis and the radius to that point. In other words, the cosine function is the horizontal component of a point on a circle. $\cos\alpha$ varies periodically in the same way as $\sin\alpha$, but $90°$ ahead. That is: $\cos\alpha$ is 1 when α is $0°$, falls to zero when $\alpha = 90°$ ($\pi/2$) and then to -1 when $\alpha = 180°$ (π), returning to zero at $\alpha = 270°$ ($3\pi/2$) and then to $+1$ again at $\alpha = 360°$ (2π). This cycle is repeated every complete revolution. The cosine function has the following properties:
$$\cos\alpha = \cos(\alpha + 360°) = \sin(\alpha + 90°)$$
$$\cos\alpha = \cos(-\alpha)$$
$$\cos(90° + \alpha) = -\cos\alpha$$
The cosine function can also be defined as an infinite series. In the range from $+1$ to -1:
$$\cos x = 1 - x^2/2! + x^4/4! - x^6/6! + \ldots$$

cosine rule In any triangle, if a, b, and c are the side lengths and γ is the angle opposite the side of length c, then
$$c^2 = a^2 + b^2 - 2ab\cos\gamma$$

cot See cotangent.

cotangent (cot) A trigonometric function of an angle equal to the reciprocal of its tangent; i.e. $\cot\alpha = 1/\tan\alpha$. *See also* trigonometry.

45

coth A hyperbolic cotangent. *See* hyperbolic functions.

coulomb Symbol: C The SI unit of electric charge, equal to the charge transported by an electric current of one ampere flowing for one second. $1 C = 1 A s$.

countable (denumerable) A set is countable if it can be put in one-one correspondence with the integers. The set of rational numbers, for example, is countable whereas the set of real numbers is not (*see* Cantor's diagonal argument). To show that the rationals are countable we need to show how they can be arranged in a series such that every rational number will be included somewhere.
If we consider the array in the diagram it is clear that every rational number will occur in it somewhere. But by starting with $1/1$ and following the path indicated we can enumerate every number in the array. If we reduce each fraction to its lowest terms and then remove any that has already occurred in the list it is clear that this will give a list in which each rational number occurs once and only once.

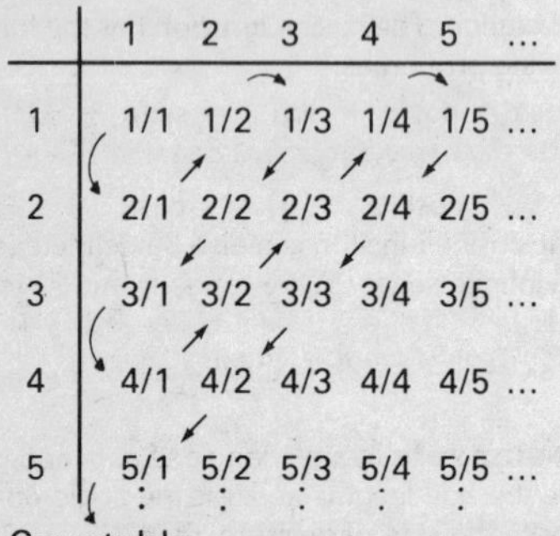

	1	2	3	4	5	...
1	1/1	1/2	1/3	1/4	1/5	...
2	2/1	2/2	2/3	2/4	2/5	...
3	3/1	3/2	3/3	3/4	3/5	...
4	4/1	4/2	4/3	4/4	4/5	...
5	5/1	5/2	5/3	5/4	5/5	...

Countable

counterclockwise *See* anticlockwise.

couple A pair of equal parallel forces in opposite directions and not acting through a single point. Their linear resultant is zero, but there is a net turning-effect (mo-

ment). The net turning effect T (the torque) is given by:
$$T = Fd_1 + Fd_2$$
F being the magnitude of each force and d_1 and d_2 the distances from any point to the lines of action of each force. This is equivalent to:
$$T = Fd$$
where d is the distance between the forces.

covariance A statistic that measures the association between two variables. If for x and y there are n pairs of values (x_i, y_i), then the covariance is defined as
$$[1/(n-1)]\Sigma(x_i - x')(y_i - y')$$
where x' and y' are the mean values.

CPU *See* central processor.

critical damping *See* damping.

critical path The sequence of operations that should be followed in order to complete a complicated process, task, etc., in the minimum time. It is usually determined by using a computer.

critical region *See* acceptance region.

cross product *See* vector product.

cross section (section) A plane cutting through a solid figure or the plane figure produced by such a cut. For example, the cross section through the middle of a sphere is a circle. A vertical cross section through an upright cone and off the axis is a hyperbola.

csc *See* cosecant.

cube 1. The third power of a number or variable. The cube of x is $x \times x \times x = x^3$ (x cubed).
2. In geometry, a solid figure that has six square faces. The volume of a cube is l^3, where l is the length of a side.

cube root An expression that has a third power equal to a given number. The cube root of 27 is 3, since $3^3 = 27$.

cubic equation A polynomial equation in which the highest power of the unknown variable is three. The general form of a cubic equation in a variable x is

$$ax^3 + bx^2 + cx + d = 0$$

where a, b, c, and d are constants. It is also sometimes written in the reduced form

$$x^3 + bx^2/a + cx/a + d/a = 0$$

In general, there are three values of x that satisfy a cubic equation. For example,

$$2x^3 - 3x^2 - 5x + 6 = 0$$

can be factorized to

$$(2x + 3)(x - 1)(x - 2) = 0$$

and its solutions (or roots) are $-3/2$, 1, and 2. On a Cartesian coordinate graph, the curve

$$y = 2x^3 - 3x^2 - 5x + 6$$

crosses the x-axis at $x = -3/2$, $x = +1$, and $x = +2$.

cuboid A box-shaped solid figure bounded by six rectangular faces. The opposite faces are congruent and parallel. At each of the eight vertices, three faces meet at right angles to each other. The volume of a cuboid is its length, l, times its breadth, b, times its height, h. The surface area is the sum of the areas of the faces, that is

$$2(l \times b) + 2(b \times h) + 2(l \times h)$$

In the special case in which $l = b = h$, all the faces are square and the cuboid is a cube of volume l^3 and surface area $6l^2$.

cumulative distribution See distribution function.

cumulative frequency The total frequency of all values up to and including the upper boundary of the class interval under consideration. See also frequency table.

curl (Symbol ∇) A vector operator on a vector function that, for a three-dimensional function, is equal to the sum of the vector product (cross product) of the unit vectors and partial derivatives in each of the component directions. That is:

$$\text{curl } F = \nabla F = \boldsymbol{i} \times \partial \boldsymbol{F}/\partial x +$$
$$\boldsymbol{j} \times \partial \boldsymbol{F}/\partial y + \boldsymbol{k} \times \partial \boldsymbol{F}/\partial z$$

where $\boldsymbol{i}$, $\boldsymbol{j}$, and $\boldsymbol{k}$ are the unit vectors in the x, y, and z directions respectively. In phys-ics, the curl of a vector arises in the relationship between electric current and magnetic flux, and in the relationship between the velocity and angular momentum of a moving fluid. See also div, grad.

curvature The rate of change of the slope of the tangent to a curve, with respect to distance along the curve. For each point on a smooth curve there is a circle that has the same tangent and the same curvature at that point. The radius of this circle, called the *radius of curvature*, is the reciprocal of the curvature, and its centre is known as the *centre of curvature*. If the graph of a function $y = f(x)$ is a continuous curve, the slope of the tangent at any point is given by the derivative dy/dx and the curvature is given by:

$$(d^2y/dx^2)/[1 + (dy/dx)^2]^{3/2}$$

curve A set of points that form or can be joined by a continuous line on a graph or other surface.

curve A set of points forming a continuous line. For example, in a graph plotted in Cartesian coordinates, the curve of the equation $y = x^2$ is a parabola. A curved surface may similarly represent a function of two variables in three-dimensional coordinates.

curvilinear integral See line integral.

cusp A sharp point formed by a discontinuity in a curve. For example, two semicircles placed side by side and touching form a cusp at which they touch.

cut See Dedekind cut.

cybernetics The branch of science concerned with control systems, especially with regard to the comparisons between those of machines and those of man and other animals. In a series of operations, information gained at one stage can be used to modify later performances of that operation. This is known as *feedback* and enables a control system to check and possibly adjust its actions when required.

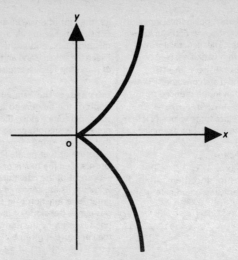

In this graph a cusp occurs at the origin o.

cycle A series of events that is regularly repeated (e.g. a single orbit, rotation, vibration, oscillation, or wave). A cycle is a complete single set of changes, starting from one point and returning to the same point in the same way.

cyclic group A group in which each element can be expressed as a power of any other element. For example, the set of all numbers that are powers of 3 could be written as { ... $3^{1/3}$, $3^{1/2}$, 3, 3^2, 3^3, ... } or { ... $9^{1/6}$, $9^{1/4}$, $9^{1/2}$, 9, $9^{3/2}$, ... }, etc. *See also* Abelian group.

cyclic polygon A polygon for which there is a circle on which all the vertices lie. All triangles are cyclic. All regular polygons are cyclic. All squares and rectangles are cyclic quadrilaterals. However, not all quadrilaterals are cyclic. Convex quadrilaterals are cyclic if the opposite angles are supplementary. For a cyclic quadrilateral with sides of length *a*, *b*, *c*, and *d* (in order) the expression (*ac* + *bd*) is equal to the product of the diagonals. This is known as *Ptolemy's theorem*.

cycloid The curve traced out by a point on a circle rolling along a straight line, for example, a point on the rim of a wheel rolling along the ground. For a circle of radius *r* along a horizontal axis, the cycloid produced is a series of continuous arcs that rises from the axis to a height 2π and fall to touch the axis again at a cusp point, where the next arc begins. The horizontal distance between successive cusps is $2\pi r$, the circumference of the circle. The length of the cycloid between adjacent cusps is 8*r*. If θ is the angle formed by the radius to a point P (*x*,*y*) on the cycloid and the radius to the point of contact with the *x*-axis, the parametric equations of the cycloid are:

$$x = r(\theta - \sin\theta)$$
$$y = r(1 - \cos\theta)$$

cylinder A solid defined by a closed plane curve (forming a base) with an identical curve parallel to it. Any line segment from a point on one curve to a corresponding point on the other curve is an *element* of the cylinder. If one of these elements moves parallel to itself round the

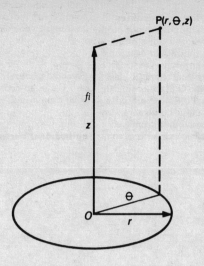

A point P(r, θ, z) in cylindrical polar coordinates.

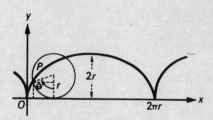

The cycloid traced out by a point P on a circle of radius r.

base it sweeps out a curved lateral surface. The line is a *generator* of the cylinder and the plane closed curve forming the base is called the *directrix*.

If the bases are circles the cylinder is a *circular cylinder*. If the bases have centres the line joining them is an axis of the cylinder. A *right cylinder* is one with its axis at right angles to the base; otherwise it is an *oblique cylinder*. The volume of a cylinder is Ah, where A is the base area and h the altitude (the perpendicular distance between the bases). For a right circular cylinder, the curved lateral surface area is $2\pi rh$, where r is the radius.

If the generator is an indefinitely extended line it sweeps out an extended surface – a *cylindrical surface*.

cylindrical helix *See* helix.

cylindrical polar coordinates

cylindrical polar coordinates A method of defining the position of a point in space by its horizontal radius r from a fixed vertical axis, the angular direction θ of the radius from an axis, and the height z above a fixed horizontal reference plane. Starting at the origin O of the coordinate system, the point P(r,θ,z) is reached by moving out along a fixed horizontal axis to a distance r, following the circumference of the horizontal circle radius r centred at O through an angle θ, and then moving vertically upwards by a distance z. For a point P(r,θ,z), the corresponding rectangular Cartesian coordinates (x,y,z) are:

$$x = r\cos\theta$$
$$y = r\sin\theta$$
$$z = z$$

Compare spherical polar coordinates. *See also* Cartesian coordinates, coordinates, polar coordinates.

cylindrical surface *See* cylinder.

D

d'Alembert's ratio test (generalized ratio test) A method of showing whether a series is convergent or divergent. The absolute value of the ratio of each term to the one before it is taken:

$$|u_{n+1}/u_n|$$

If the limit of this is l as n tends to infinity and l is less than 1, then the series is convergent. If l is greater than 1, the series is divergent. If l is equal to 1, the test fails and some other method has to be used. *See also* limit.

damped oscillation An oscillation with an amplitude that progressively decreases with time. *See* damping.

damping The reduction in amplitude of a vibration with time by some form of resistance. A swinging pendulum will at last come to rest; a plucked string will not vibrate for long – in both cases internal and/or external resistive forces progressively reduce the amplitude and bring the system to equilibrium.

In many cases the damping force(s) will be proportional to the object's speed. In any event, energy must be transferred from the vibrating system to overcome the resistance. Where damping is an asset (as in bringing the pointer of a measuring instrument to rest), the optimum situation occurs when the motion comes to zero in the shortest time possible, without vibration. This is *critical damping*. If the resistive force is such that the time taken is longer than this, *overdamping* occurs. Conversely, *underdamping* involves a longer time with vibrations of decreasing amplitude.

data (now often used as a singular noun) The facts that refer to or describe an object, idea, condition, situation, etc. In computing, data can be regarded as the facts on which a program operates as opposed to the instructions in the program. It can only be accepted and processed by the computer in binary form. Data is some-times considered to be numerical information only. *See also* program.

data bank A large collection of organized computer data, from which particular pieces of information can be readily extracted. *See also* database.

database A large collection of organized data providing a common pool of information for users, say, in the various sections of a large organization. Information can be added, deleted, and updated as required. The management of a database is very complicated and costly so that computer programs have been developed for this purpose. These programs allow the information to be extracted in many different ways. For example, a request could be put in for an alphabetical list of men over a certain age and living in a specified area, in which their employment and income should be given. Alternatively the request could be for an alphabetical list of men over a certain age and income level in which their address and form of employment should be given.

data processing The sequence of operations performed on data in order to extract information or to achieve some form of order. The term usually means the processing of data by computers but can also include its observation and collection.

debug To detect, locate, and correct errors or faults (bugs) that occur in computer programs or in pieces of computer equipment. Since programs and equipment are often highly complicated, debugging can be a tedious and lengthy job. Programming errors may result from the incorrect coding of an instruction (known as a *syntax error*) or from using instructions that will not give the required solution to a problem (a *logic error*). Syntax errors can usually be detected and located by the compiler; logic errors can be more difficult to find. *See also* program.

deca-

deca- Symbol: da A prefix denoting 10. For example, 1 decametre (dam) = 10 metres (m).

decagon A plane figure with ten straight sides. A *regular decagon* has ten equal sides and ten equal angles of 36°.

decahedron A polyhedron that has ten faces. *See* polyhedron.

deci- Symbol: d A prefix denoting 10^{-1}. For example, 1 decimetre (dm) = 10^{-1} metre (m).

decibel Symbol: dB A unit of power level, usually of a sound wave or electrical signal, measured on a logarithmic scale. The threshold of hearing is taken as 0 dB in sound measurement. Ten times this power level is 10 dB. The fundamental unit is the *bel*, but the decibel is almost exclusively used (1 dB = 0.1 bel).
A power P has a power level in decibels given by:

$$10 \log_{10}(P/P_0)$$

where P_0 is the reference power.

decimal Denoting or based on the number ten. The numbers in common use for counting form a decimal number system. A *decimal fraction* is a rational number written as units, tenths, hundredths, thousandths, and so on. For example, $\frac{1}{4}$ is 0.24 in decimal notation. This type of decimal fraction (or decimal) is a *finite decimal* because the third and subsequent digits after the decimal point are 0. Some rational numbers, such as 5/27 (= 0.185 185 185 ...) cannot be written as an exact decimal, but result in a number of digits that repeat indefinitely. These are called *repeating decimals*. All rational numbers can be written as either finite decimals or repeating decimals. A decimal that is not finite and does not repeat is an irrational number and can be quoted to any number of decimal places, but never exactly. For example, π to an accuracy of six decimal places is 3.141 593 and to seven decimal places is 3.141 592 7.
A decimal measure is any measuring system in which larger and smaller units are

derived by multiplying and dividing the basic unit by powers of ten. *See also* metric system.

decision box *See* flowchart.

decomposition 1. The process of breaking a fraction up into partial fractions.
2. Decomposition of a vector v is the process of writing it in the form

$$v = \text{grad}(\phi) + \text{curl}(A)$$

where ϕ is a scalar and A is a vector. Every vector may be decomposed in this form.

Dedekind cut A method of defining the real numbers, starting from the rational numbers. A Dedekind cut is a division of the rational numbers into two disjoint sets, A and B, which are nonempty and which satisfy the conditions: (1) if $x \in A$ and $y \in B$ then $x < y$, and (2) A has no largest member (or, equivalently, B has no smallest member). The real numbers can be defined as the set of all Dedekind cuts and can be shown to have all the requisite properties.

deduction A series of logical steps in which a conclusion is reached directly from a set of initial statements (premisses). A deduction is valid if a sentence or statement that asserts the premisses and denies the conclusion is a contradiction. *Compare* induction. *See* contradiction.

definite integral (Riemann integral) The result of integrating any function of a single variable, f(x), between two specified values of x: x_1 and x_2. The definite integral of f(x) is written

$$\int_{x_1}^{x_2} f(x)dx$$

If the general expression for the integral of f(x) (its indefinite integral) is another function of x, g(x), the definite integral is given by:

$$g(x_1) - g(x_2)$$

Compare indefinite integral. *See also* integration.

52

definition In a measurement, the accuracy with which the instrument reading reflects the true value of the quantity being measured. *See also* accuracy.

deformation A geometrical transformation that stretches, shrinks, or twists a shape but does not break up any of its lines or surfaces. It is often called, more precisely, a *continuous deformation*. *See also* topology, transformation.

degenerate conic *See* conic.

degree 1. Symbol: ° A unit of plane angle equal to one ninetieth of a right angle. **2.** A unit of temperature. *See* Celsius degree, Fahrenheit degree, kelvin. **3.** The exponent of a variable. For instance $3x^3$ has a degree of 3. If there are several terms, the sum of the exponents is used; $5xy^2z^3$ has a degree of 6. **4.** The highest power of a variable in a polynomial. For example in $x^3 + 2x + 1$, the degree is 3. **5.** The highest power of an equation. For example, the degree of the equation $x^4 + 2x = 0$ is 4. **6.** The highest power to which the highest order derivative is raised in a differential equation. For example, the degree of
$$(d^2y/dx^2)^3 + dy/dx = 0$$
is three. The degree of
$$d^3y/dx^3 + 2y(d^2y/dx^2)^2 = 0$$
is one. *See also* differential equation.

De l'Hôpital's rule The rule stating that the limit of the ratio of two functions of the same variable (x) as x approaches a value a, is equal to the limit of the ratio of their derivatives with respect to x. That is, the limit of $f(x)/g(x)$ as $x{\to}a$ is the limit of $f'(x)/g'(x)$ as $x{\to}a$.
De l'Hôpital's rule can be used to find the limits of $f(x)/g(x)$ at points at which both $f(x)$ and $g(x)$ are zero and the ratio is therefore indeterminate. Any function that gives rise to an indeterminate form and that can be expressed as a ratio of two functions, can be dealt with in this way. For example, in
$$F(x) = (x^2 - 3)/(x - 3)$$
writing

$$f(x) = (x^2 - 3)$$
and
$$g(x) = (x - 3)$$
gives
$$F(x) = f(x)/g(x)$$
The limit of $F(x)$ as $x{\to}3$ is indeterminate (since $x - 3 = 0$). It can be obtained by using the limit of
$$f'(x)/g'(x) = 2x$$
as $x{\to}3$. Thus the limit is 6.
If $f'(x)/g'(x)$ also gives an indeterminate form at $x = a$, De l'Hôpital's rule can be applied again, differentiating as many times as is necessary.

De Moivre's theorem A formula for calculating a power of a complex number. If the number is in the polar form
$$z = r(\cos\theta + i\sin\theta)$$
then $z^n = r^n(\cos n\theta + i\sin n\theta)$

De Morgan's laws Two laws governing the relation between complementation, intersection, and union of sets. If $\bar{A}$ represents the complement of the set A (i.e. the set of all things not in A) then De Morgan's laws state that: (1) $(\overline{A \cup B}) = \bar{A} \cap \bar{B}$; and (2) $(\overline{A \cap B}) = \bar{A} \cup \bar{B}$. Analogous laws are true for any finite intersection or union of sets. Parallel laws exist in other areas, e.g. in propositional logic the equivalences ${\sim}(p \,\&\, q) \equiv {\sim}p \lor {\sim}q$ and ${\sim}(p \lor q) \equiv {\sim}p \,\&\, {\sim}q$ are also known as De Morgan's laws.

denominator The bottom part of a fraction. For example, in the fraction $^3/_4$, 4 is the denominator and 3 is the numerator. The denominator is the divisor.

dense A set S is said to be dense in another set T if every point of T either belongs to S or is a limit point of S. If it is obvious what the other set is, then a set is sometimes simply said to be dense; e.g. if we are considering sets of real numbers. For example, the set of rational numbers is dense (on the real line) because every point of the real line is either a rational number or is the limit of a sequence of rational numbers. Another way of putting this is to say that any interval on the real line will always contain infinitely many rationals.

density The amount of matter per unit volume; mass divided by volume.

denumerable set A set in which the elements can be counted. For example, the set of prime numbers, although infinite, can be counted, as can the set of positive even integers. These are known as denumerably infinite sets. The set of rational numbers, on the other hand, is not denumerable because between any two elements there can always be found a third. *See* countable. *See also* set.

dependent 1. An equation is *dependent* on a set of equations if *every* set of values of the unknowns that satisfy the set of equations also satisfies this equation. Otherwise the equation is *independent* of the set of equations.
2. Two events are *independent* if the occurrence or nonoccurrence of one of them does not affect the probability of the occurrence of the other. Otherwise the events are *dependent*. If A and B are events whose probabilities are $P(A)$ and $P(B)$, then A and B are independent if and only if $P(A \text{ and } B) = P(A)P(B)$. For example, each toss of a coin is an independent event.
3. A set of functions $\{f_1, \ldots f_n\}$ are *dependent* if one can be expressed as a function of the others or equivalently there exists an expression $F(f_1, \ldots f_n) \equiv 0$ with not all $\partial F/\partial f_i = 0$. Otherwise they are *independent*. For example, the functions $2x + y$ and $4x + 2y + 6$ are dependent since $4x + 2y + 6 = 2(2x + y) + 6$.
4. *See* variable.
5. A set of vectors, matrices, or other objects $\{x_1, \ldots x_n\}$ is said to be *linearly dependent* if there exists a linear relation
$$a_1x_1 + a_2x_2 + \ldots + a_nx_n = 0$$
with at least one of the coefficients nonzero. A set of objects is *linearly independent* if it is not linearly dependent. It should be noted that the dependence is relative to the set from which we may pick the coefficients $a_1, \ldots a_n$. For example, 2 and π are linearly independent with re-

spect to the rational numbers but linearly dependent with respect to the real numbers. This is the case since a relation of the form $a_1 2 + a_2 \pi = 0$ does not exist if a_1 and a_2 are rational numbers but if a_1 and a_2 are allowed to be irrational numbers we may take $a_1 = \pi$, $a_2 = 2$.

dependent variable *See* variable.

deposit A sum of money paid by a buyer, either to reserve goods or property that he wishes to buy at a later date or as the first of a series of instalments in a hire-purchase agreement. If the buyer fails to complete the purchase the deposit is normally forfeited.

depth The distance downwards from a reference level or backwards from a reference plane. For example, the distance below a water surface and the distance between a wall surface and the back of an alcove in the wall, are both called depths.

derivative The result of differentiation. *See* differentiation.

derived unit A unit defined in terms of base units, and not directly from a standard value of the quantity it measures. For example, the newton is a unit of force defined as a kilogram metre seconds^{-2} (kg m s^{-2}). *See also* SI units.

determinant A function of a square matrix derived by multiplying and adding the elements together to obtain a single number. For example, in a 2×2 matrix the determinant is $a_1b_2 - a_2b_1$. This is written as a square array in vertical lines, symbol D_2, and is called a *second order determinant*. Determinants occur in simultaneous equations. The solution of
$$a_1x + b_1y + c_1 = 0$$
and
$$a_2x + b_2y + c_2 = 0$$
is
$$x = (b_1c_2 - c_1b_2)/D_2$$
and
$$y = (c_1a_2 - a_1c_2)/D_2$$
If a_1, a_2, b_1, b_2, c_1, and c_2 are 1, 2, 3, 4, 5, and 6 respectively, then $D_2 = -2$ and

$$\begin{vmatrix} a_1 & b_1 \\ a_2 & b_2 \end{vmatrix} = a_1b_2 - a_2b_1$$

The second order determinant of a 2×2 matrix.

$$\begin{vmatrix} a_1 & b_1 & c_1 \\ a_2 & b_2 & c_2 \\ a_3 & b_3 & c_3 \end{vmatrix} = a_1b_2c_3 - a_1b_3c_2 + a_2b_3c_1 - a_2b_1c_3 \\ + a_3b_1c_2 - a_3b_2c_1$$

The third order determinant of a 3×3 matrix.

$$\begin{vmatrix} a_1 & b_1 & c_1 \\ a_2 & b_2 & c_2 \\ a_3 & b_3 & c_3 \end{vmatrix} = a_1 \begin{vmatrix} b_2 & c_2 \\ b_3 & c_3 \end{vmatrix} - b_1 \begin{vmatrix} a_2 & c_2 \\ a_3 & c_3 \end{vmatrix} + c_1 \begin{vmatrix} a_2 & b_2 \\ a_3 & b_3 \end{vmatrix}$$

$$= a_1a_1' - b_1b_1' + c_1c_1'$$
$$= a_1a_1' - a_2a_2' + a_3a_3'$$

$$\begin{array}{ccc} + & - & + \\ - & + & - \\ + & - & + \end{array}$$

A third order determinant is equal to the sum along any row, or down any column, of the product of each element with its cofactor. The cofactors are given alternate positive and negative signs in the pattern shown. Fourth and higher order determinants can be calculated in a similar way.

$x = [(3 \times 6) - (5 \times 4)]/-2 = 1$
and

$y = [(5 \times 2) - (1 \times 6)]/-2 = -2$
A *third order determinant* has three rows and columns and arises in a similar way in sets of three simultaneous equations in three variables.
The determinant of a transpose of a matrix, $|\tilde{A}|$, is equal to the determinant of the matrix, $|A|$. If the position of any of the rows or columns in the matrix is changed, the determinant remains the same.

developable surface A surface that can be rolled out flat onto a plane. The lateral surface of a cone, for example, is developable. A spherical surface is not.

deviation *See* mean deviation, standard deviation.

diagonal Joining opposite corners. A diagonal of a square, for example, cuts it into two congruent right angled triangles. In a solid figure, usually a polyhedron, a diagonal plane is one that passes through two edges that are not adjacent.

diagonal matrix A square matrix in which all the elements are zero except those on the leading diagonal, that is, the first element in the first row, the second element in the second row, and so on. Diagonal matrices, unlike most others, are commutative in matrix multiplication.

$$\begin{pmatrix} a_{11} & 0 & 0 \\ 0 & a_{22} & 0 \\ 0 & 0 & a_{33} \end{pmatrix}$$

A 3 × 3 diagonal matrix.

diameter The distance across a plane figure or a solid at its widest point. The diameter of a circle or a sphere is twice the radius.

diametral Denoting a line or plane that forms a diameter of a figure. For example, a cross section through the centre of a sphere is a diametral plane.

dichotomy, principle of In logic, the principle that a proposition is either true or false, but not both. For example, for two numbers x and y either $x = y$ or $x \neq y$, but not both.

difference The result of subtracting one quantity or expression from another.

differential An infinitesimal change in a function of one or more variables, resulting from a small change in the variables. For example, if f(x) is a function of x, and f changes by Δf as a result of a change Δx in x, the differential df, is defined as the limit of Δf as Δx becomes infinitely small. That is, df = f'(x)dx, where f'(x) is the derivative of f with respect to x. This is a *total differential*, because it takes into account changes in all of the variables, just one in this case.

For a function of two variables, f(x,y) the rate of change of f with respect to x is the partial derivative ∂f/∂x. The change in f resulting from changing x by dx and keeping y constant is the *partial differential*, (∂f/∂y).dy. For any function, the total differential is the sum of all the partial differentials. For f(x,y):

$$df = (\partial f/\partial x).dx + (\partial f/\partial y).dy$$

See also differentiation.

differential equation An equation that contains derivatives. An example of a simple differential equation is:

$$dy/dx + 4x + 6 = 0$$

To solve such equations it is necessary to use integration. The equation above can be rearranged to give:

$$dy = -(4x + 6)dx$$

integrating both sides:

$$\int dy = \int -(4x + 6)dx$$

which gives:

$$y = -2x^2 - 6x + C$$

where C is a constant of integration. The value of C can be found if particular values of x and y are known: for example, if $y = 1$ when $x = 0$ then $C = 1$, and the full solution is

$$y = -2x^2 - 6x + 1$$

Note that the solution to a differential equation is itself an equation. Differentiating the solution gives the original equation. Equations like that above, which contain only first derivatives (dy/dx) are said to be *first order*, if they contain second derivatives they are *second order*; in general, the *order* of a differential equation is the highest derivative in the equation. The *degree* of a differential equation is the highest power of the highest order derivative.

The differential equation in the example given is a first order and first degree equation. It is an example of a type of equation solvable by separating the variables onto both sides of the equation, so that each can be integrated (the *variables separable* method of solution). Another type of first-order first-degree equation is one of the form:

$$dy/dx = f(y/x)$$

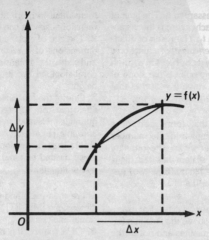

Differentiation of a function $y = f(x)$. The derivative dy/dx is the limit of $\Delta y/\Delta x$ as Δx and Δy become infinitely small.

Such equations are known as homogeneous differential equations. An example is the equation:

$$dy/dx = (x^2 + y^2)/x^2$$

To solve homogeneous equations a substitution is made, $y = mx$, where m is a function of x. Then:

$$dy/dx = m + x\,dm/dx$$

and

$$(x^2 + y^2)/x^2 = (x^2 + m^2x^2)/x^2$$

So the equation becomes:

$$m + x\,dm/dx = (x^2 + m^2x^2)/x^2$$

or:

$$x\,dm/dx = 1 + m^2 - m$$

The equation can now be solved by separating the variables.

An equation of the form:

$$dy/dx + P(x)y = Q(x)$$

where $P(x)$ and $Q(x)$ are functions of x (but not y), is a *linear differential* equation. Equations of this type can be put in a solvable form by multiplying both sides by the expression:

$$\exp(\textstyle\int P(x)dx)$$

This is known as an *integrating factor*. For example, the differential equation

$$dy/dx + y/x = x^2$$

is a linear first-order differential equation. The function $P(x)$ is $1/x$, so the integrating factor is:

$$\exp(\textstyle\int dx/x)$$

which is $\exp(\log x)$; i.e. x. Multiplying both sides of the equation by x gives:

$$x\,dy/dx + y = x^3$$

The left-hand side of the equation is equal to $d(xy)/dx$, so the equation becomes

$$d(xy)/dx = x^3$$

Integrating both sides gives:

$$xy = x^4/4 + C$$

where C is a constant.

differentiation A process for finding the rate at which one variable quantity changes with respect to another. For example, a car might travel along a road from position x_1 to position x_2 in a time interval t_1 to t_2. Its average velocity is $(x_2 - x_1)/(t_2 - t_1)$, which can be written $\Delta x/\Delta t$, where Δx represents the change in x in the time Δt. However, the car might accelerate or decelerate in this interval and it may be necessary to know the velocity at a particular instant, say t_1. In this case the time interval Δt is made infinitely small, i.e. t_2 can

57

...ry to t_1. The limit of ...es zero is the instan-... . The result of differ-...*rivative*) of a function y .../dx or f'(x). On a graph ...any point is the slope of ...he curve $y = f(x)$ at that ...integration.

digit A symbol that forms part of a number. For example, in the number 3121 there are four digits. The ordinary (decimal) number system has ten digits (0–9), whereas the binary (base two) system needs only two, 0 and 1.

digital Using numerical digits. For example, a digital watch shows the time in numbers of hours and minutes and not as the position of hands on a dial. In general, digital devices work by some kind of counting process, either mechanical or electronic. The abacus is a very simple example. Early calculating machines counted with mechanical relays. Modern calculators use electronic switching circuits.

digital computer *See* computer.

dihedral Formed by two intersecting planes. Two planes intersect along a straight line (edge). The *dihedral angle* (or *dihedron*) between the planes is the angle between two lines (one in each plane) drawn perpendicular to the edge from a point on the edge. The dihedral angle of a polyhedron is the angle between two faces.

dihedron *See* dihedral.

dilatation A geometrical mapping or projection in which a figure is 'stretched', not necessarily by the same amount in each direction. A square, for example, may be mapped into a rectangle by dilatation, or a cube into a cuboid.

dimension 1. The number of coordinates needed to represent the points on a line, shape, or solid. A plane figure is said to be two-dimensional; a solid is three-dimensional. In more abstract studies n-dimensional spaces can be used.
2. The size of a plane figure or solid. The dimensions of a rectangle are its length and width; the dimensions of a rectangular parallelopiped are its length, width, and height.
3. One of the fundamental physical quantities that can be used to express other quantities. Usually, mass [M], length [L], and time [T] are chosen. Velocity, for example, has dimensions of $[L][T]^{-1}$ (distance divided by time). Force, as defined by the equation:

$$F = ma$$

where m is mass and a acceleration, has dimensions $[M][L][T]^{-2}$. *See also* dimensional analysis.
4. Of a matrix, the number of rows or number of columns. A matrix with 4 rows and 5 columns is 4×5 matrix.

dimensional analysis The use of the dimensions of physical quantities to check relationships between them. For instance, Einstein's equation $E = mc^2$ can be checked. The dimensions of speed^2 are $([L][T]^{-1})^2$, i.e. $[L]^2[T]^{-2}$, so mc^2 has dimensions of $[M][L]^2[T]^{-2}$. Energy also has these dimensions since it is force $[M][L][T]^{-2}$ multiplied by distance [L]. Dimensional analysis is also used to obtain the units of a quantity and to suggest new equations.

directed Having a specified positive or negative sign, or a definite direction. A directed number usually has one of the signs + or − written in front of it. A directed angle is measured from one specified line to the other. If the direction were reversed, the size of the angle would be a negative number.

direction A property of vector quantities, usually defined in reference to a fixed origin and axes. The direction of a curve at a point is the angle from the x axis to the tangent at the point.

Diophantine equation *See* indeterminate equation.

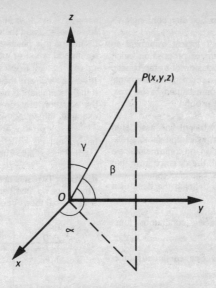

The direction angles α, β, and γ, that the line *OP* makes with the *x*, *y*, and *z* axes respectively in a three-dimensional Cartesian coordinate system

directional derivative The rate of change of a function with respect to distance *s* in a particular direction, or along a specified curve. Going from a point P(x,y,z) in the direction that makes angles α, β, and γ with the *x*, *y*, and *z* axes respectively, the directional derivative of a function f(x,y,z) is

$$df/ds = (\partial f/\partial x)\cos\alpha + (\partial f/\partial y)\cos\beta + (\partial f/\partial z)\cos\gamma$$

If there is a direction for which the directional derivative is a maximum, then this derivative is the gradient of f (grad f) at point P. *See also* grad.

direction angle The angle between a line and one of the axes in a rectangular Cartesian coordinate system. In a plane system, it is the angle, α, that the line makes with the positive direction of the *x*-axis. In three dimensions, there are three

direction angles, α, β, and γ, for the *x*, *y*, and *z* axes respectively. If two direction angles are known, the third can be calculated by the relationship:

$$\cos^2\alpha + \cos^2\beta + \cos^2\gamma = 1$$

Cosα, cosβ, and cosγ are called the *direction cosines* of the line, sometimes given the symbols *l*, *m*, and *n*. Any three numbers in the ratio *l*:*m*:*n* are called the *direction numbers* or the *direction ratio* of the line. If a line joins the point A (x_1,y_1,z_1) and the point B (x_2,y_2,z_2) and the distance between A and B is *D*, then

$$l = (x_2 - x_1)/D$$
$$m = (y_2 - y_1)/D$$
$$n = (z_2 - z_1)/D$$

direction cosines *See* direction angle.

direction numbers *See* direction angle.

direction ratio *See* direction angle.

direct proof A logical argument in which the theorem or proposition being proved is the conclusion of a step-by-step process based on a set of initial statements that are known or assumed to be true. *Compare* indirect proof.

directrix 1. A straight line associated with a conic, from which the shortest distance to any point on the conic maintains a constant ratio with the distance from that point to the focus. *See also* conic.
2. A plane curve defining the base of a cone or cylinder.

discontinuity *See* continuous function.

discontinuous *See* continuous function.

discount 1. The difference between the issue price of a stock or share and its nominal value when the issue price is less than the nominal value. *Compare* premium.
2. A reduction in the price of an article or commodity for payment in cash (*cash discount*), or for a large order (*bulk discount*), or for a retailer who will be selling the goods on to members of the public (*trade discount*).

discrete Denoting a set of events or numbers in which there are no intermediate levels. The set of integers, for example, is discrete but the set of rational numbers is not. Between any two rational numbers, no matter how close, there can always be found another rational number. The results of tossing dice form a discrete set of events, since a die has to land on one of its six faces. Putting the shot, on the other hand, does not have a discrete set of outcomes, since it may travel for any distance in a continuous range of lengths.

discriminant The expression $(b^2 - 4ac)$ in a quadratic equation of the form $ax^2 + bx + c = 0$. If the roots of the equation are equal, the discriminant is zero. For example, in

$$x^2 - 4x + 4 = 0$$

$b^2 - 4ac = 0$ and the only root is 2. If the discriminant is positive, the roots are different and real. For example, in

$$x^2 + x - 6 = 0$$

$b^2 - 4ac = 25$ and the roots are 2 and -3. If the discriminant is negative, the roots of the equation are complex numbers. For example, the equation:

$$x^2 + x + 1 = 0$$

has roots $[-\frac{1}{2} - (\sqrt{3}/2]i$ and $[-\frac{1}{2} + (\sqrt{3}/2]i$. *See also* quadratic equation.

disjoint Two sets are said to be disjoint if they have no members in common; i.e. if $A \cap B = 0$ then A and B are disjoint.

P	Q	$P \lor Q$
T	T	T
T	F	T
F	T	T
F	F	F

disjunction (inclusive)

disjunction Symbol: $\lor$ In logic, the relationship *or* between two propositions or statements. Disjunction can be either inclusive or exclusive. *Inclusive disjunction* (sometimes called *alternation*) is the one most commonly used in mathematical logic, and can be interpreted as 'one or the other or both'. For two propositions P and Q, $P \lor Q$ is false if P and Q are both false, and true in all other cases. The more rarely used *exclusive disjunction* can be interpreted as 'either one or the other but not both'. With this definition $P \lor Q$ is false when P and Q are both true, as well as when they are both false. The truth table definitions for both types of disjunction are shown in the illustration. *Compare* conjunction. *See also* truth table.

disk A device that is widely used in computer systems to store information. It is a flat circular metal plate coated usually on both sides with a magnetizable substance. Information is stored in the form of small magnetized spots, which are closely packed in concentric *tracks* on the coated

P	Q	P∨Q
T	T	F
T	F	T
F	T	T
F	F	F

disjunction (exclusive)

$$\sum_{}^{n} |x_j - x| / n$$

If values $X_1, X_2, \ldots, X_k$ occur with frequencies $f_1, f_2, \ldots f_k$ it becomes

$$\sum_{}^{n} f_j |x_j - x| / \sum f_j$$

See also mean.

surfaces of the disk. The spots are magnetized in one of two directions so that the information is in binary form. The magnetization pattern of a group of spots represents a letter, digit (0–9), or some other character. One disk can store several million characters. Information can be altered or deleted as necessary by magnetic means. Disks are usually stacked on a common spindle in a single unit known as a *disk pack*. Disk packs storing 200 million characters are common.

Information can be recorded on a disk using a special typewriter; this method is known as *key-to-disk*. The information is fed into a computer using a complex device called a *disk unit*. The disk pack is rotated at very great speed in the disk unit. Small electromagnets, known as *read–write heads*, move radially in and out over the surfaces of the rotating disks. They extract (read) or record (write) items of information at specified locations on a track, following instructions from the central processor. The time to reach a specified location is very short. This factor, together with the immense storage capacity, makes the disk unit a major backing store in a computer system. *Compare* drum, magnetic tape. *See also* floppy disk.

diskette *See* floppy disk.

dispersion A measure of the extent to which data are spread about an average. The range, the difference between the largest and smallest results, is one measure. If P_r is the value below which $r\%$ of the results occur, then the range can be written as $(P_{100} - P_0)$. The interquartile range is $(P_{75} - P_{25})$. The semi-interquartile range is $(P_{75} - P_{25})/2$. The mean deviation of $X_1, X_2, \ldots, X_n$ measures the spread about the mean X and is

displacement Symbol: *s* The vector form of distance, measured in metres (m) and involving direction as well as magnitude.

dissipation The removal of energy from a system to overcome some form of resistive force. Without resistance (as in motion in a vacuum) there can be no dissipation. Dissipated energy normally appears as thermal energy.

distance Symbol: *d* The length of the path between two points. The SI unit is the metre (m). Distance may or may not be measured in a straight line. It is a scalar; the vector form is displacement.

distance formula The formula for the distance between two points (x_1, y_1) and (x_2, y_2) in Cartesian coordinates. It is:
$$\sqrt{[(x_1 - x_2)^2 + (y_1 - y_2)^2]}$$

distance ratio (velocity ratio) For a machine, the ratio of the distance moved by the effort in a given time to the distance moved by the load in the same time. *See also* machine.

distribution function For a random variable *x*, the function f(*x*) that is equal to the probability of each value of *x* occurring. If all values of *x* between *a* and *b* are equally likely, *x* has a *uniform distribution* in this interval and a graph of the distribution function f(*x*) against *x* is a horizontal line. For example, the probability of the results 1 to 6 when throwing dice is a uniform distribution. Continuous random variables usually have a varying distribution function with a maximum value x_m and in which the probability of *x* decreases as *x* moves away from x_m. The *cumulative* *distribution* function F(x) is the probability of a value less than or equal to x. For the dice example, F(x) is a step function that increases from zero to one in six equal steps. For continuous functions, F(x) is often an s-shaped curve. In both cases F(x) is the area under the curve of f(x) to the left of x.

distributive Denoting an operation that is independent of being carried out before or after another operation. For two operations · and ∘, · is distributive with respect to ∘ if $a·(b∘c) = (a·b)∘(a·c)$ for all values of a, b, and c. In ordinary arithmetic, multiplication is distributive with respect to addition $[a(b + c) = ab + ac]$ and to subtraction.
In set theory intersection (∩) is distributive with respect to union (∪):
$$[A∩(B∪C) = (A∩B)∪(A∩C)]$$
See also associative, commutative.

div (divergence) Symbol: ∇. A scalar operator that, for a three-dimensional vector function F(x,y,z), is the sum of the scalar products of the unit vectors and the partial derivatives in each of the three component directions. That is:
$$div\ F = ∇.F = \textbf{\textit{i}}.∂F/∂x + \textbf{\textit{j}}.∂F/∂y + \textbf{\textit{k}}.∂F/∂z$$
In physics, div F is used to describe the excess flux leaving an element of volume in space. This may be a flow of liquid, a flow of heat in a field of varying temperature, or an electric or magnetic flux in an electric or magnetic field. If there is no source of flux (heat source, electric charge, etc.) within the volume, then div F = 0 and the total flux entering the volume equals the total flux leaving. *See also* grad.

divergent sequence A sequence in which the difference between the nth term and the one after it is constant or increases as n increases. $\{1,2,4,8, \dots \}$ is divergent. A divergent sequence has no limit. *Compare* convergent sequence. *See also* divergent series, geometric sequence, sequence.

divergent series A series in which the sum of all the terms after the nth term

does not decrease as n increases. A divergent series, unlike a convergent series, has no sum to infinity. An infinite series $a_1 + a_2 + \dots$ is *divergent* if the partial sums $a_1 + a_2 + \dots + a_n$ tend to infinity as n tends to infinity. For example, the series $1 + 2 + 3 + 4 + \dots$ is divergent. *Compare* convergent series. *See also* divergent sequence, geometric series, series.

dividend 1. The number into which another number (the divisor) is divided to give a quotient. For example, in $16 ÷ 3$, 16 is the dividend and 3 is the divisor.
2. A share of the profits of a limited company paid to shareholders. The rate of dividend paid by a company will depend on its profits in the preceding year. It is expressed as a percentage of the nominal value of the shares. For example, a 10% dividend on a 75p share will pay 7.5p share (independent of the market price of the share). *See also* yield.

dividers A drawing instrument, similar to compasses, but with sharp points on both ends. Dividers are used for measuring lengths on a drawing, or for dividing straight lines.

division Symbol: ÷ The binary operation of finding the quotient of two quantities. Division is the inverse operation to multiplication. In arithmetic, the division of two numbers is not commutative ($2 ÷ 3 ≠ 3 ÷ 2$), nor associative $[(2 ÷ 3) ÷ 4 ≠ 2 ÷ (3 ÷ 4)]$. The identity element for division is one only when it comes on the right hand side ($5 ÷ 1 = 5$ but $1 ÷ 5 ≠ 5$). *Compare* multiplication.

divisor The number by which another number (the dividend) is divided to give a quotient. For example, in $16 ÷ 3$, 16 is the dividend and 3 is the divisor. *See also* factor.

documentation Written instructions and comments that give a full description of a computer program. The documentation describes the purposes for which the program can be used, how it operates, the exact form of the inputs and outputs, and how the computer must be operated. It allows the program to be amended when necessary or to be converted for use on different types of machines.

domain A set of numbers or quantities on which a mapping is, or may be, carried out. In algebra, the domain of the function $f(x)$ is the set of values that the independent variable x can take. If, for example, $f(x)$ represents taking the square root of x, then the domain might be defined as all the positive rational numbers. *See also* range.

D operator The differential operator d/dx. The derivative df/dx of a function $f(x)$ is often written as Df. This notation is used in solving differential equations. A second derivative, d^2f/dx^2, is written as D^2f, a third derivative, d^3f/dx^3, as D^3f, and so on. In some ways, the D operator can be treated like an ordinary algebraic quantity, despite the fact that it has no numerical value. For example, the differential equation
$$d^2y/dx^2 + 2xdy/dx + dy/dx + 2x = 0$$
or
$$D^2y + 2xD + D + 2x = 0$$
can be factorized to $(D + 2x)(D + 1) = 0$. $(D + 2x)$ then operates on the function $(D + 1)$. *See also* differential equations.

dot product *See* scalar product.

double-angle formulae *See* addition formulae.

double integral The result of integrating the same function twice, first with respect to one variable, holding a second variable constant, and then with respect to the second variable, holding the first variable constant. For example, if $f(x,y)$ is a function of the variables x and y, then the double integral, first with respect to x and then with respect to y, is:
$$\iint f(x,y)dydx$$
This is equivalent to summing $f(x,y)$ over intervals of both x and y, or to finding the volume bounded by the surface representing $f(x,y)$. The integral is not affected by the order in which the integrations are carried out if they are definite integrals. An-

other kind of double integral is the result of integrating twice with respect to the same variable. For example, if a car's acceleration a increases with time t in a known way, then the integral
$$\int adt$$
is the velocity (v) expressed as a function of time; the double integral
$$\iint adt^2 = \int v.dt = x$$
where x is the distance travelled as a function of time.

double point A singular point on a curve at which the curve crosses itself or is tangential to itself. There are several types of double point. At a *node* the curve crosses over itself forming a loop. In this case it has two distinct tangents. At a *cusp* it double back on itself and has only one tangent. At an *acnode* two arcs of curve touch each other and have the same tangent but, unlike a cusp, the arcs continue through the singular point to form four arms. An *isolated double point* may also occur. This satisfies the equation of the curve but does not lie on the main arc of the curve. *See also* isolated point, multiple point.

drum A metal cylinder coated with a magnetizable substance and used in a computer system to store information. The information is stored in the form of small magnetized spots, which are closely packed in concentric *tracks* around the circumference of the drum. When in use the drum is rotated at high speed. Small electromagnets, called *read−write heads*, are fixed in position over each track. They extract (read) or record (write) items of information in particular locations on the track, as specified by the central processor. The time taken to obtain an item of information is extremely short. Drums are now used only for a few special applications in computing. *Compare* disk, magnetic tape. *See also* central processor, random access.

duality The principle in mathematics whereby one true theorem can be obtained from another merely by substituting certain words in a systematic way. In a

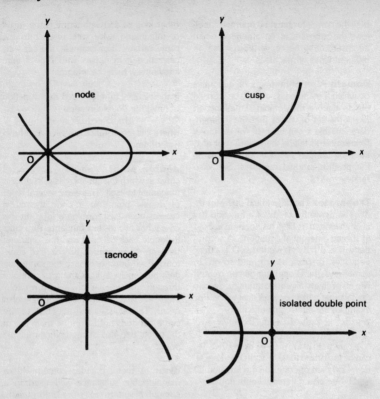

Four types of double point at the origin of a two-dimensional Cartesian coordinate system.

plane, 'point' and 'line' are *dual elements* and, for example, drawing a line through a point and marking a point on a line are *dual operations*. Theorems that can be obtained from one another by replacing each element and operation by its dual are *dual theorems*. In logic, 'implied' and 'is implied by' are *dual relations* and may be interchanged together with the logical connectives 'and' and 'or'. In set theory, the relations 'is contained in' and 'contains' are dual relations and can be interchanged along with the 'union' and 'intersection'. In general, the principle of duality is found where the structure under consideration is a lattice. *See* lattice.

dummy variable A symbol that can be replaced by any other symbol without changing the meaning of the expression in which it occurs.

duodecahedron A polyhedron that has twelve faces. A *regular duodecahedron* has twelve congruent faces, each one being a regular pentagon. *See also* polyhedron.

duodecimal Denoting or based on twelve. In a duodecimal number system there are twelve different digits instead of ten. If, for example, ten and eleven were given the symbols *A* and *B* respectively, 12 would be written as 10 and 22 as 1A. Duodecimal numbers are of little use, but some duodecimal units (1 foot = 12 inches) are still in use. *Compare* binary, decimal, hexadecimal, octal.

duty A tax levied on certain kinds of transactions. Examples include taxes on importing and exporting goods and the tax levied on alcohol and tobacco products.

dynamic friction *See* friction.

dynamics The study of how objects move under the action of forces. This includes the relation between force and motion and the description of motion itself. *See also* mechanics.

dyne Symbol: dyn The former unit of force used in the c.g.s. system. It is equal to 10^{-5} N.

E

e A fundamental mathematical constant defined by the series:
$$e = 1 + 1/1! + 1/2! \ldots + 1/n! + \ldots$$
e is the base of natural logarithms. It is both irrational and transcendental. It is related to π by the formula $e^{i}r = -1$. The function e^x has the property that it is its own derivative, that is, $de^x/dx = e^x$.

eccentric Denoting intersecting circles, spheres, etc., that do not have the same centre. *Compare* concentric.

eccentricity A measure of the shape of a conic. The eccentricity is the ratio of the distance of a point on the curve from a fixed point (the focus) to the distance from a fixed line (the directrix). For a parabola, the eccentricity is 1. For a hyperbola, it is greater than 1, and for an ellipse, it is between 0 and 1. A circle has an eccentricity of 0.

ecliptic The apparent path along which the Sun moves each year. It is the great circle formed by the intersection of the plane of the Earth's orbit with the celestial sphere.

edge A straight line where two faces of a solid meet. A cube has twelve edges.

efficiency Symbol: η A measure used for processes of energy transfer; the ratio of the useful energy produced by a system or device to the energy input. For example, the efficiency of an electric motor is the ratio of its mechanical power output to the electrical power input. There is no unit of efficiency; however efficiency is often quoted as a percentage. In practical systems some dissipation of energy always occurs (by friction, air resistance, etc.) and the efficiency is less than 1. For a machine, the efficiency is the force ratio divided by the distance ratio.

effort The force applied to a machine. *See* machine.

eigenfunction *See* eigenvalue.

eigenvalue (from German *eigen* = 'allowed') An eigenvalue for a linear transformation L on a vector space V is a scalar λ for which there is a nonzero solution vector $\boldsymbol{v}$ in V such that $L\boldsymbol{v} = \lambda\boldsymbol{v}$ and $\boldsymbol{v}$ satisfies any given boundary conditions. The vector $\boldsymbol{v}$ is an *eigenvector* (or *characteristic vector*) belonging to the eigenvalue λ. An eigenvector for a linear operator on a vector space whose vectors are functions is also called an *eigenfunction*. As an example, consider the equation $-y'' = \lambda y$ with boundary conditions $y = 0$ when $x = a$ or b. A nonzero solution exists only if $\lambda = n^2\pi^2$ where n is an integer. The solution then is
$$y = C[\sin n\pi(x - a)]/(b - a)$$
where C is an arbitrary constant. The eigenvalues of the equation are given by $n^2\pi^2$ (where n is an integer). The corresponding eigenfunctions (or eigenvectors) are given by
$$y = C[\sin(x - a)]/(b - a)$$

eigenvector *See* eigenvalue.

elastic collision A collision for which the restitution coefficient is equal to one. Kinetic energy is conserved during an elastic collision. In practice, collisions are not perfectly elastic as some energy is transferred to internal energy of the bodies. *See also* restitution, coefficient of.

electronvolt Symbol: eV A unit of energy equal to $1.602\ 191\ 7 \times 10^{-19}$ joule. It is defined as the energy required to move an electron charge across a potential difference of one volt. It is normally used only to measure energies of elementary particles, ions, or states.

element 1. A single item that belongs to, or is a member of, a set. 'February', for example, is an element of the set {month in the year}. The number 5 is an element

$$a_1x + b_1y + c_1z = 0$$
$$a_2x + b_2y + c_2z = 0$$
$$a_3x + b_3y + c_3z = 0$$

$$\begin{vmatrix} a_1 & b_1 & c_1 \\ a_2 & b_2 & c_2 \\ a_3 & b_3 & c_3 \end{vmatrix} = 0$$

A set of three simultaneous equations and the eliminant given by the corresponding matrix determinant equation.

of the set of integers between 2 and 10. In set notation this is written as

$$5 \in \{2,3,4,5,6,7,8,9,10\}$$

2. A line segment forming part of the curved surface of a surface, as of a cone or cylinder.
3. A small part of a line, surface, or volume summed by integration.
4. (of a matrix) See matrix.

eliminant (characteristic, resultant) The relationship between coefficients that results from eliminating the variable from a set of simultaneous equations. For example, in the equations

$$a_1x + b_1y + c_1 = 0$$
$$a_2x + b_2y + c_2 = 0$$
$$a_3x + b_3y + c_3 = 0$$

the eliminant is given by:

$$\begin{vmatrix} a_1 & b_1 & c_1 \\ a_2 & b_2 & c_2 \\ a_3 & b_3 & c_3 \end{vmatrix} = 0$$

elimination Removing one of the unknowns in an algebraic equation, for example, by the substitution of variables or by cancellation.

ellipse A conic with an eccentricity between 0 and 1. An ellipse has two foci. A line through the foci cuts the ellipse at two *vertices*. The line segment between the vertices is the *major axis* of the ellipse. The point on the major axis mid-way between the vertices is the *centre* of the ellipse. A line segment through the centre perpendicular to the major axis is the *minor axis*. Either of the chords of the ellipse through a focus parallel to the minor axis is a *latus rectum*. The area of an ellipse is πab, where a is half the major axis and b is half the minor axis. (Note that for a circle, in which the eccentricity is zero, $a = b = r$ and the area is πr^2.)

The *sum property* of an ellipse is that for any point on the ellipse the sum of the distances from the point to each focus is a constant. The ellipse also has a reflection property; for a given point on the ellipse the two lines from each focus to the point make equal angles with a tangent at that point.

In Cartesian coordinates the equation:

$$x^2/a^2 + y^2/b^2 = 1$$

represents an ellipse with its centre at the origin. The major axis is on the x-axis and the minor axis on the y-axis. The major axis is $2a$ and the minor axis is $2b$. The foci of the ellipse are at the points $(+ea,0)$ and $(-ea,0)$, where e is the eccentricity. The two directrices are the lines $x = a/e$ and $x = -a/e$. The length of the latus rectum is $2b^2/a$. See also conic.

ellipsoid A solid body or curved surface in which every plane cross-section is an ellipse or a circle. An ellipsoid has three axes of symmetry. In three-dimensional Cartesian coordinates, the equation of an ellipsoid with its centre at the origin is:

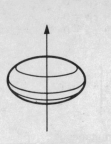 

prolate ellipsoid   oblate ellipsoid

An ellipsoid can be generated by rotating an ellipse about one of its axes. An oblate ellipsoid is generated by rotation about the major axis and a prolate ellipsoid is generated by rotation about a minor axis.

$$x^2/a^2 + y^2/b^2 + z^2/c^2 = 1$$

where a, b, and c are the points at which it crosses the x, y, and z axes respectively. In this case the axes of symmetry are the coordinate axes. A *prolate ellipsoid* is one generated by rotating an ellipse about its major axis. An *oblate ellipsoid* is generated by rotation about the minor axis.

empirical Derived directly from experimental results or observations.

empty set (null set) Symbol: Ø The set that contains no elements. For example, the set of 'natural numbers less than 0' is an empty set. This could be written as $\{m: m \in N; m<0\} = \emptyset$.

energy Symbol: W A property of a system – its capacity to do work. Energy and work have the same unit: the joule (J). It is convenient to divide energy into kinetic energy (energy of motion) and potential energy ('stored' energy). Names are given to many different forms of energy (chemical, electrical, nuclear, etc.); the only real difference lies in the system under discussion. For example, chemical energy is the kinetic and potential energies of electrons in a chemical compound. *See also* kinetic energy, potential energy, mass–energy equation.

enlargement A geometrical projection that produces an image larger (smaller if the scale factor is less than 1) than, but similar to, the original shape. *See also* projection.

entailment In logic, the relationship that holds between two (or more) propositions when one can be deduced from the other. If conclusion C is deducible from premises A and B, then A and B are said to *entail* C. *See* deduction.

enumerable *See* countable.

envelope Consider a one-parameter family of curves in three dimensions – i.e. a family of curves that can be represented in terms of a common parameter that is constant along each curve, but is changed from curve to curve. The envelope of this family of curves is the surface traced out by these curves. This surface is tangent to every curve of the family. Its equation is obtained by eliminating the parameter between the equation of the curve and the partial derivative of this equation with respect to the parameter. For example, the

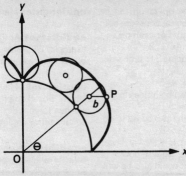

The epicycloid traced out by a point
P on a circle of radius *b* that rolls
round a large circle of radius *a*.

envelope of the family of paraboloids given by $x^2 + y^2 = 4a(z - a)$ is the equation obtained by eliminating *a* from the equations $x^2 + y^2 = 4a(z - a)$ and $z - 2a = 0$, i.e. the circular cone $x^2 + y^2 = z^2$.

epicycle A circle that rolls around the circumference of another, tracing out an epicycloid. *See* epicycloid.

epicycloid The plane curve traced out by a point on a circle or *epicycle* rolling along the outside of another fixed circle. For example, if a small cog wheel turns on a larger stationary wheel, then a point on the rim of the smaller wheel traces out an epicycloid. In a two-dimensional Cartesian-coordinate system that has a fixed circle of radius *a* centred at the origin and another of radius *b* rolling around the circumference, the epicycloid is a series of continuous arcs that move away from the first circle to a distance 2*b* and then return to touch it again at a cusp point where the next arc begins. The epicycloid has only one arc if $a = b$, two if $a = b/2$, and so on. If the angle between the radius from the origin to the moving point of contact between the two circles is θ, the epicycloid is defined by the parametric equations:
$x = (a + b) \cos\theta - a \cos[(a + b)\theta/a]$
$y = (a + b) \sin\theta - a \sin[(a + b)\theta/a]$

equality Symbol: = The relationship between two quantities that have the same value or values. If two quantities are not equal, the symbol $\neq$ is used. For example, $x \neq 0$ means that the variable *x* cannot take the value zero. When the equality is only approximate, the symbol $\simeq$ is used. For example, if Δx is small compared to *x* then $x + (\Delta x)^2 \simeq x$. When two expressions are exactly equivalent the symbol $\equiv$ is used. For example $\sin^2\alpha \equiv 1 - \cos^2\alpha$ because it applies for all values of the variable α. *See also* equation, inequality.

equation A mathematical statement that one expression is equal to another, that is, two quantities joined by an equals sign. An algebraic equation contains unknown or variable quantities. It may state that two quantities are identical for all values of the variables, and in this case the identity symbol $\equiv$ is normally used. For example:

$$x^2 - 4 \equiv (x - 2)(x + 2)$$

is an *identity* because it is true for all values that *x* might have. The other kind of algebraic equation is a *conditional equation*, which is true only for certain values of the variables. To solve such an equation, that is, to find the values of the variables at which it is valid, it often has to be rearranged into a simpler form. In simplifying an equation, the same operation can be carried out on the expressions on both sides of the equals sign. For example,

$$2x - 3 = 4x + 2$$

can be simplified by adding 3 to both sides to give

$$2x = 4x + 5$$

then subtracting $4x$ from both sides to give

$$-2x = 5$$

and finally dividing both sides by -2 to obtain the solution $x = -5/2$.

This kind of equation is called a linear equation, because the highest power of the variable x is one. It could also be written in the form $-2x - 5 = 0$. On a Cartesian coordinate graph,

$$y = -2x - 5$$

is a straight line that crosses the x-axis at $x = -5/2$.

Performing the same operation on both sides of an equation does not necessarily give an equation exactly equivalent to the original. For example, starting with $x = y$ and squaring both sides gives $x^2 = y^2$, which means that $x = y$ or $x = -y$. In this case the symbol $\Rightarrow$ is used between the equations, meaning that the first implies the second, but the second does not imply the first. That is,

$$x = y \Rightarrow x^2 = y^2$$

Where the two equations are equivalent, the symbol $\Leftrightarrow$ is used, for example,

$$2x = 2 \Leftrightarrow x = 1$$

equations of motion Equations that describe the motion of an object with constant accleration (a). They relate the velocity v_1 of the object at the start of timing to its velocity v_2 at some later time t and to the object's displacement s. They are:

$$v_2 = v_1 + at^2$$
$$s = (v_1 + v_2)t/2$$
$$s = v_1 t + at^2/2$$
$$s = v_2 t - at^2/2$$
$$v_2{}^2 = v_1{}^2 + 2as$$

equator On the Earth's surface, the circle formed by the plane cross-section that perpendicularly bisects the axis of rotation. The plane in which the circle lies is called the *equatorial plane*. A similar circle on any sphere with a defined axis is also called an equator, or equatorial circle.

equiangular Having equal angles.

equidistant At the same distance. For example, all points on the circumference of a circle are equidistant from the centre.

equilateral Having sides of equal length. For example, an equilateral triangle has three sides of equal length (and equal interior angles of 60°).

equilibrant A single force that is able to balance a given set of forces and thus cause equilibrium. It is equal and opposite to the resultant of the given forces.

equilibrium A state of constant momentum. An object is in equilibrium if:
(1) its linear momentum does not change (it moves in a straight line at constant speed and has constant mass, or is at rest);
(2) its angular momentum does not change (its rotation is zero or constant).
For these conditions to be met:
(1) the resultant of all outside forces acting on the object must be zero (or there are no outside forces);
(2) there is no resultant turning-effect (moment).
An object is not in equilibrium if any of the following are true:
(1) its mass is changing;
(2) its speed is changing;
(3) its direction is changing;
(4) its rotational speed is changing.
See also stability.

equivalence *See* biconditional.

equivalence principle *See* relativity, theory of.

equivalence relation A binary relation R defined on a set S is said to be an equivalence relation if it satisfies the following three properties: (1) xRx for every x in S – this is the property of *reflexivity*, (2) if xRy then yRx – this is the property of *symmetry*, and (3) if xRy and yRz then xRz – this is the property of *transitivity*. Such relations are especially important because they partition the set on which they are defined into disjoint classes, known as *equivalence classes*.

Eratosthenes, sieve of A method of finding prime numbers. To find all the prime numbers less than a given number n one first goes through all the numbers from 2 to n removing all those that are multiples of 2. Then all those after 3 are examined and all the multiples of 3 are removed. One proceeds in this way with all the numbers less than or equal to $\sqrt{n}$. Only prime numbers will remain.

erg A former unit of energy used in the c.g.s. system. It is equal to 10^{-7} joule.

error The uncertainty in a measurement or estimate of a quantity. For example, on a mercury thermometer, it is often possible to read temperature only to the nearest degree Celsius. A temperature of 20°C should then be written as (20 ± 0.5)°C because it really means 'between 19.5°C and 20.5°C'. There are two basic types of error. *Random error* occurs in any direction, cannot be predicted, and cannot be compensated for. It includes the limitations in the accuracy of the measuring instrument and the limitations in reading it. *Systematic error* arises from faults or changes in conditions that can be corrected for. For example, if a 1 gram weight used on a balance is 2 milligrams underweight, every measurement taken with it will be 2 milligrams less than the correct value.

escape speed (escape velocity) The minimum initial speed (velocity) that an object must have in order to escape from the surface of a planet (or moon) against the gravitational attraction. The escape speed is equal to $\sqrt{(2GM/r)}$, where G is the gravitational constant, M is the mass of the planet, and r is the radius of the planet. The concept also applies to the escape of the object from a distant orbit.

estimate A rough calculation, usually involving one or more approximations, made to give a preliminary answer to a problem.

ether (aether) A hypothetical fluid, formerly thought to permeate all space and to be the medium through which electromagnetic waves were propagated. *See* relativity, theory of.

Euclidean algorithm A method of finding the highest common factor of two positive integers. The smaller number is divided into the larger. The remainder is then divided into the smaller number, obtaining a second remainder. This second remainder is then divided into the first remainder, to give a third remainder. This is divided into the second, and so on, until a zero remainder is obtained. The remainder preceding this is the highest common factor of the two numbers. For example, the numbers 54 and 930. Dividing 54 into 930 gives 17 with a remainder of 12. Dividing 12 into 54 gives 4 with a remainder of 6. Dividing 6 into 12 gives 2 with a remainder of 0. Thus 6 is the highest common factor of 54 and 930.

Euclidean geometry A system of geometry described by the Greek mathematician Euclid in his book *Elements* (*c.* 300 BC). It is based on a number of definitions – point, line, etc. – together with a number of basic assumptions. These were axioms or 'common notions' – for example, that the whole is greater than the part – and postulates about geometric properties – for example, that a straight line is determined by two points. Using these basic ideas a large number of theorems were proved using formal deductive arguments. The basic assumptions of Euclid have been modified, but the system is essentially that used today for 'pure' geometry.
One important postulate in Euclid's system is that concerned with parallel lines (the *parallel postulate*). Its modern form is that if a point lies outside a straight line only one straight line can be drawn through that point parallel to the other line. *See* non-Euclidean geometry.

Euler characteristic A topological property of a curve or surface. For a curve, the Euler characteristic is the number of vertices minus the number of closed continuous line segments between. For example, any polygon has an Euler characteristic of zero. For a surface, the Euler characteristic is equal to the number of vertices plus the number of faces minus the number of edges. For example, a cube has an Euler characteristic of 2, and a cylinder, a Möbius strip, and a Klein bottle have an Euler characteristic of zero.

Euler's constant A fundamental mathematical constant defined by the limit of:
$$1 + 1/2 + 1/3 + \ldots + 1/n - \log n$$
as $n \rightarrow \infty$. To six figures its value is 0.577 216. It is not known whether Euler's constant is irrational or not.

Euler's formula 1. (for polyhedra) The formula that relates the number of vertices v, faces f, and edges e in a polyhedron, that is:
$$v + f - e = 2$$
For example, a cube has eight vertices six faces and twelve edges:
$$8 + 6 - 12 = 2$$
Using the theorem it can be shown that there are only five regular polyhedrons.
2. The definition of the function $e^i g$ for any real value of θ, where i is the square root of -1, is
$$e^i g = \cos\theta + i \sin\theta$$
Any complex number $z = x + iy$ can be written in this form. $x = r\cos\theta$ and $y = r\sin\theta$ are real, with r and θ representing z on an Argand diagram. Note that putting $\theta = \pi$ gives $e^i r = -1$ and $\theta = 2\pi$ gives $e^2 r^i = 1$.

even Divisible by two. The set of even numbers is $\{2,4,6,8, \ldots \}$. *Compare* odd.

even function A function f(x) of a variable x, for which $f(-x) = f(x)$. For example, $\cos x$ and x^2 are even functions of x. *Compare* odd function.

event In probability, an *event* is any subset of the possible outcomes of an experiment. The event is said to *occur* if the outcome is a member of the subset. For example, if two dice are thrown, an event is a subset of all ordered pairs (m,n) where

m and n are each one of the integers 1, 2, 3, 4, 5, 6. Thus $\{(1,3),(2,2),(3,1)\}$ is an event, which may also be described as 'obtaining a sum of four'. *See also* dependent.

evolute The evolute of a given curve is the locus of the centres of curvature of all the points on the curve. The evolute of a surface is another surface formed by the locus of all the centres of curvature of the first surface.

excluded middle, law of *See* laws of thought.

exclusive disjunction (exclusive OR) *See* disjunction.

exclusive OR gate *See* logic gate.

existence theorem A theorem that proves that one or more mathematical entities of a certain kind exists; e.g. that a function has a zero or a fixed point. An existence proof may be indirect and show that a certain entity must exist without giving any information about it or how to find it.

existential quantifier In mathematical logic, a symbol meaning 'there is (are)'. It is usually written $\exists$. The quantifier is followed by a variable that it is said to *bind*. Thus $(\exists x)F(x)$ means 'There is something that has property F'.

expansion A quantity expressed as a sum of a series of terms. For instance, the expression:
$$(x + 1)(x + 2)$$
can be expanded to:
$$x^2 + 3x + 2$$
Often a function can be written as an infinite series that is convergent. The function can then be approximated to any required accuracy by taking the sum of a sufficient number of terms at the beginning of the series. There are general formulae for expanding some types of expression. For example, the expansion of $(1 + x)^n$ is
$$1 + nx + [n(n - 1)/2!]x^2 + [n(n - 1)(n - 2)/3!] x^3 + \ldots$$

where x is a variable between -1 and $+1$, and n is an integer. *See* binomial expansion, determinant, Fourier series, Taylor series.

expectation *See* expected value.

expected value (expectation) The value of a variable quantity that is calculated to be most likely to occur. If x can take any of the set of discrete values $\{x_1, x_2, \ldots x_n\}$, which have corresponding probabilities $\{p_1, p_2, \ldots p_n\}$, then the expected value is
$$E(x) = x_1 p_1 + x_2 p_2 + \ldots + x_n p_n$$
If x is a continuous variable with a probability density function $f(x)$, then
$$E(x) = \int_{-\infty}^{\infty} x f(x) dx$$

explicit Denoting a function that contains no dependent variables. *Compare* implicit.

exponent A number or symbol placed as a superscript after an expression to indicate the power to which it is raised. For example, x is an exponent in y^x and in $(ay + b)^x$.
The laws of exponents are used for combining exponents of numbers as follows:
Multiplication:
$$x^a x^b = x^{a+b}$$
Division:
$$x^a / x^b = x^{a-b}$$
Power of a power:
$$(x^a)^b = x^{ab}$$
Negative exponent:
$$x^{-a} = 1/x^a$$
Fractional exponent:
$$x^{a/b} = \sqrt[b]{x^a}$$
A number raised to the power zero is equal to 1; i.e. $x^0 = 1$.

exponential A function or quantity that varies as the power of another quantity. In $y = 4^x$, y is said to vary exponentially with respect to x. The function e^x (or expx), where e is the base of natural logarithms, is the exponential of x.
The infinite series

$$1 + x + x^2/2! + x^3/3! + \ldots + x^n/n! + \ldots$$
is equal to e^x and is known as the *exponential series*. The exponential form of a complex number is
$$re^{ig} = r(\cos\theta + i\sin\theta)$$
See also complex number, Euler's formula, Taylor series.

exponential series The infinite power series that is the expansion of the function e^x, namely:
$$1 + x + x^2/2! + x^3/3! + \ldots + x^n/n! + \ldots$$
This series is convergent for all real-number values of the variable x.
Replacing x by $-x$ gives an alternating series for e^{-x}:
$$1 - x + x^2/2! - x^3/3! \ldots$$
Series for sinhx and coshx can be obtained by combining series for e^x and e^{-x}.

expression A combination of symbols (representing numbers of other mathematical entities) and operations; e.g. $3x^2$, $\sqrt{(x^2 + 2)}$, $e^x - 1$.

exterior angle The angle formed on the outside of a plane figure between the extension of one straight edge beyond a vertex, and the outer side of the other straight edge at that vertex. In a triangle, the exterior angle at one vertex equals the sum of the angles on the insides of the other two vertices, i.e. the sum of the interior opposite angles. *Compare* interior angle.

extraction The process of finding a root of a number.

extrapolation The process of estimating the value of a function or quantity outside a known range of values. For example, if the speed of an engine is controlled by a lever, and depressing the lever by two, four, and six centimetres gives speeds of 20, 30, and 40 revolutions per second respectively, then one can extrapolate from this information and make the assumption that depressing it by a further two centimetres will increase the speed to 50 revolutions per second. Extrapolation

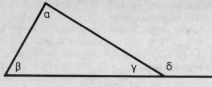

The exterior angle $\delta = 180° - \gamma = \alpha + \beta$.

can also be carried out graphically; for example, a graph can be drawn over a known range of values and the resulting curve extended. The further from the known range this line is taken, the greater will be the uncertainty in the extrapolation. The case in which the graph of the behaviour is a straight line (as in the example above) is a *linear extrapolation*. *Compare* interpolation.

F

face A flat surface on the outside of a solid figure. A cube has six identical faces.

factor (divisor) A number by which another number is divided. *See also* common factor.

factorial The product of all the whole numbers less than or equal to a number. For example, factorial 7, written 7!, is equal to $7 \times 6 \times 5 \times 4 \times 3 \times 2 \times 1$. Factorial zero is defined as 1.

factorization The process of changing algebraic or numerical expressions from a sum of terms into a product. For example, the left side of the equation $4x^2 - 4x - 8 = 0$ can be factorized to $(2x + 2)(2x - 4)$ making it easy to solve for x. As the product of the two factors is 0 when either of the factors is 0, it follows that $(2x + 2) = 0$ and $(2x - 4) = 0$ will provide solutions, i.e. $x = -1$ and $x = 2$.

factor theorem The condition that $(x - a)$ is a factor of a polynomial f(x) in a variable x if and only if f(a) = 0. For example, if f(x) = $x^2 + x - 6$, f(2) = 4 + 2 − 6 = 0 and f(−3) = 9 − 3 − 6 = 0, so the factors of f(x) are $(x - 2)$ and $(x + 3)$. The factor theorem is derived from the remainder theorem.

Fahrenheit degree Symbol: °F A unit of temperature difference equal to one hundred and eightieth of the difference between the temperatures of freezing and boiling water. On the Fahrenheit scale water freezes at 32°F and boils at 212°F. To convert from a temperature on the Fahrenheit scale (T_F) to a temperature on the Celsius scale (T_C) the following formula is used: $T_F = 9T_C/5 + 32$.

fallacy *See* logic.

family A set of related curves or figures. For example, the equation $y = 3x + c$ represents a family of parallel straight lines.

farad Symbol: F The SI unit of capacitance. When the plates of a capacitor are charged by one coulomb and there is a potential difference of one volt between them, then the capacitor has a capacitance of one farad. $1 F = 1 CV^{-1}$, 1 farad = 1 coulomb per volt.

fathom A unit of length used to measure depth of water. It is equal to 6 feet (1. 8288 m).

F distribution The statistical distribution followed by the ratio of variances, s_1^2/s_2^2, of pairs of random samples, size n_1 and n_2, taken from a normal distribution. It is used to compare different estimates of the same variance.

feedback *See* cybernetics.

femto- Symbol: f A prefix denoting 10^{-15}. For example, 1 femtometre (fm) = 10^{-15} metre (m).

Fermat's last theorem The theorem that the equation
$$x^n + y^n = z^n$$
where n is an integer greater than 2, can have no solution for x, y, and z. Fermat wrote in the margin of a book on equations that he had discovered a 'truly wonderful' proof of the theorem but that the margin was too small to write it down. Unfortunately, he died before he could supply the proof, and no one since has been able to prove that the theorem is correct, although no solution has yet been found.

fermi A unit of length equal to 10^{-15} metre. It was formerly used in atomic and nuclear physics.

Fibonacci numbers The infinite sequence in which successive numbers are formed by adding the two previous numbers, that is:

$$1, 1, 2, 3, 5, 8, 13, 21, \ldots$$

fictitious force A force in a system that arises because of the frame of reference of the observer. Such 'forces' are said to be 'fictitious' because they do not actually exist; they can be removed by transfer to a different frame of refernce. Examples are centrifugal force and Coriolis force.

field 1. A set of entities with two operations, called addition and multiplication. The entities form a commutative group under addition with 0 as the identity element. If 0 is omitted, the entities form a commutative group under multiplication. Also the distribution law, $a(b + c) = ab + ac$, applies for all a, b, and c. An example of a field is the set of rational numbers.
2. A region in which a particle or body exerts a force on another particle or body through space. In a gravitational field a mass is supposed to affect the properties of the surrounding space so that another mass in this region experiences a force. The region is thus called a 'field of force'. Electric, magnetic, and electromagnetic fields can be similarly described. The concept of a field was introduced to explain action at a distance.

figure A shape formed by a combination of points, lines, curves, or surfaces. Circles, squares, and triangles are plane figures. Spheres, cubes, and pyramids are solid figures.

finite decimal *See* decimal.

finite sequence *See* sequence.

finite series *See* series.

finite set A set that has a fixed countable number of elements. For example, the set of 'months in the year' has 12 members and is therefore a finite set. *Compare* infinite set.

first order differential equation A differential equation in which the highest derivative of the dependent variable is a first derivative. *See* differential equation.

fixed-point theorem A theorem that demonstrates that a function leaves one point in its domain unchanged, i.e. for which $f(x) = x$. One celebrated example is *Brouwer's fixed point theorem*, which states that any continuous transformation of a circular disc onto itself must have a fixed point.

flip-flop *See* bistable circuit.

floating objects, law of *See* flotation, law of.

floppy disk (diskette) A device that can be used to store information, consisting of a flexible plastic disk with a magnetic coating on one or both sides. It is permanently encased in a stiff envelope inside which it can be made to rotate. A read–write head operates through a slot in the envelope. *See* disk.

flotation, law of An object floating in a fluid displaces its own weight of fluid. This follows from Archimedes' principle for the special case of floating objects. (A floating object is in equilibrium, its only support coming from the fluid. It may be totally or partly submerged.)

flowchart A diagram on which can be represented the major steps in a process used, say, in industry, or a problem to be investigated, or a task to be performed. A flowchart is built up from a number of boxes connected by arrowed lines. The boxes, of various shapes, have a lable attached showing for example the operation or calculation to be done at each step. At a *decision box* a question is asked. The answer, either yes or no, determines which of two possible paths to take. Computer programs are often written by first drawing a flowchart of the problem or task in hand. *See also* program.

fluid ounce *See* ounce.

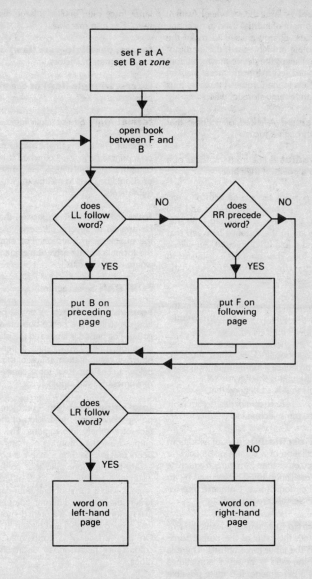

A flow chart for finding which page a word is on in this dictionary (assuming that it is in). F is a front marker, B a back marker, LL is the first word on a left-hand page, RR the last word on a right-hand page, and LR the first word on the right-hand page.

flywheel A large heavy wheel (with a large moment of inertia) used in mechanical devices. Energy is used to make the wheel rotate at high speed; the inertia of the wheel keeps the device moving at constant speed, even though there may be fluctuations in the torque. A flywheel thus acts as an 'energy-storage' device.

focal chord A chord of a conic that passes through a focus.

focal radius A line from the focus of a conic to a point on the conic.

focus A point associated with a conic. The distance between the focus and any point on the curve is in a fixed ratio (the eccentricity) to the distance between that point and a line (the directrix). An ellipse has two foci. The sum of the distances to each focus is the same for all points on the ellipse. *See also* conic.

foot Symbol: ft The unit of length in the f.p.s. system (one third of a yard). It is equal to 0.304 8 metre.

force Symbol: F That which tends to change an object's momentum. Force is a vector; the unit is the newton (N).
In SI, this unit is so defined that:
$$F = d(mv)/dt$$
from Newton's second law.

forced oscillation (forced vibration) The oscillation of a system or object at a frequency other than its natural frequency. Forced oscillation must be induced by an external periodic force. *Compare* free oscillation. *See also* resonance.

force ratio (mechanical advantage) For a machine, the ratio of the output force (load) to the input force (effort). There is no unit; the ratio is, however, sometimes given as a percentage. It is quite possible for force ratios far greater than one to be obtained. Indeed many machines are designed for this so that a small effort can overcome a large load. However the efficiency cannot be greater than one and a large force ratio implies a large distance ratio. *See also* machine.

forces, parallelogram (law) of *See* parallelogram of vectors.

forces, triangle (law) of *See* triangle of vectors.

formal logic *See* symbolic logic.

format The arrangement of information on a printed page, on a punched card, in a computer storage device, etc., that must or should be used to meet with certain requirements.

formula A general expression that can be applied to several different values of the quantities in question. For example, the formula for the area of a circle is πr^2, where r is the radius.

FORTRAN *See* program.

Foucault pendulum A simple pendulum consisting of a heavy bob on a long string. The period is large and the plane of vibration rotates slowly over a period of time as a result of the rotation of the Earth below it. The apparent force causing this movement is the Coriolis force.

four-colour problem A problem in topology concerning the division of the surface of a sphere into regions. The name comes from the colouring of maps. It appears that in colouring a map it is not necessary to use more than four colours to distinguish regions from each other. Two regions with a common line boundary between them need different colours, but two regions meeting at a point do not. This was proved by Appel and Haken in 1976. On the surface of a torus only seven colours are necessary to distinguish regions.

Fourier series A method of expanding a function by expressing it as an infinite series of periodic functions (sines and cosines). The frequencies of the sines and cosines are increased by a constant factor with each successive term. The general mathematical form of the Fourier series is:

$$f(x) = a_0/2 + (a_1\cos x + b_1\sin x) +$$
$$(a_2\cos 2x + b_2\sin 2x) +$$
$$(a_3\cos 3x + b_3\sin 3x) +$$
$$(a_n\cos nx + b_n\sin nx) + \ldots$$

The constants a_0, a_1, b_1, etc., called *Fourier coefficients*, are obtained by the formulae:

$$a_0 = (1/\pi)\int_{-\pi}^{\pi} f(x)dx$$

$$a_n = (1/\pi)\int_{-\pi}^{\pi} f(x)\cos nx\,dx$$

$$b_n = (1/\pi)\int_{-\pi}^{\pi} f(x)\sin nx\,dx$$

f.p.s. system A system of units that uses the foot, the pound, and the second as the base units. It has now been largely replaced by SI units for scientific and technical work although it is still used to some extent in the USA.

fraction A number written as a quotient; i.e. as one number divided by another. For example, in the fraction $2/3$, 2 is known as the *numerator* and 3 as the *denominator*. When both numerator and denominator are integers, the fraction is known as a *simple*, *common*, or *vulgar fraction*. A *complex fraction* has another fraction as numerator or denominator, for example $(2/3)/(5/7)$ is a complex fraction. A *unit fraction* is a fraction that has 1 as the numerator. If the numerator of a fraction is less than the denominator, the fraction is known as a *proper fraction*. If not, it is an *improper fraction*. For example, $5/2$ is an improper fraction and can be written as $2\frac{1}{2}$. In this form it is called a *mixed number*.

In adding or subtracting fractions, the fractions are put in terms of their lowest common denominator. For example,

$$1/2 + 1/3 = 3/6 + 2/6 = 5/6$$

In multiplying fractions, the numerators are multiplied and the denominators multiplied. For example:

$$2/3 \times 5/7 = (2 \times 5)/(3 \times 7) = 10/21$$

In dividing fractions, one fraction is inverted thus:

$$2/3 \div 1/2 = 2/3 \times 2/1$$

See also ratio.

frame of reference A set of coordinate axes with which the position of any object may be specified as it changes with time. The origin of the axes and their spatial directions must be specified at every instant of time for the frame to be fully determined.

freedom, degrees of The number of independent quantities that are necessary to determine an object or system. The number of degrees of freedom is reduced by constraints on the system since the number of independent quantities necessary to determine the system is also reduced. For example, a point in space has three degrees of freedom since three coordinates are needed to determine its position. If the point is constrained to lie on a curve in space, it has then only one degree of freedom, since only one parameter is needed to specify its position on the curve.

free oscillation (free vibration) An oscillation at the natural frequency of the system or object. For example, a pendulum can be forced to swing at any frequency by applying a periodic external force, but it will swing freely at only one frequency, which depends on its length and mass. *Compare* forced oscillation. *See also* resonance.

free variable In mathematical logic, a variable that is not within the scope of any quantifier. (If it is within the scope of a quantifier it is *bound*.) In the formula $F(y) \rightarrow (\exists x)G(x)$, y is a free variable.

French curve A drawing instrument consisting of a rigid piece of plastic or metal sheeting with curved edges. The curvature of the edges varies from being almost straight to very tight curves, so that a part of the edge can be chosen to guide a pen or pencil along any desired curvature. Another instrument that serves the same purpose consists of a deformable strip of lead

French curve

bar cut into short sections and surrounded by a thick layer of plastic. This is bent to form any required curvature.

frequency Symbol: f, ν The number of cycles per unit time of an oscillation (e.g. a pendulum, vibrating system, wave, alternating current, etc.). The unit is the hertz (Hz). The symbol f is used for frequency, although ν is often employed for the frequency of light or other electromagnetic radiation.
Angular frequency (ω) is related to frequency by $\omega = 2\pi f$.

frequency curve A smoothed frequency polygon for data that can take a continuous set of values. As the amount of data is increased and the size of class interval decreased, the frequency polygon more closely approximates a smooth curve. Relative frequency curves are smoothed relative frequency polygons. *See also* skewness, frequency polygon.

frequency function 1. The function that gives the values of the frequency of each result or observation in an experiment. For a large sample that is representative of the whole population, the observed frequency function will be the same as the probability distribution function f(x) of a population variable x. *See also* distribution function.
2. *See* random variable.

frequency polygon The graph obtained when the mid-points of the tops of the rectangles in a histogram with equal class intervals are joined by line segments.

The area under the polygon is equal to the total area of the rectangles. *See also* histogram.

frequency table A table showing how often each type (class) of result occurs in a sample or experiment. For example, the weekly wages received by 100 employees in a company could be shown as the number in each range from £50.00 to £74.99, £75.00 to £99.99, and so on. In this case the representative value of each class (the *class mark*) is £$(50 + 74.99)/2$. etc. *See also* histogram.

friction A force opposing the relative motion of two surfaces in contact. In fact, each surface applies a force on the other in the opposite direction to the relative motion; the forces are parallel to the line of contact. The exact causes of friction are still not fully understood. It probably results from minute surface roughness, even on apparently 'smooth' surfaces. Frictional forces do not depend on the area of contact. Presumably lubricants act by separating the surfaces. For friction between two solid surfaces, *sliding friction* (or *kinetic friction*) opposes friction between two moving surfaces. It is less than the force of *static* (or *limiting*) *friction*, which opposes slip between surfaces that are at rest. *Rolling friction* occurs when a body is rolling on a surface: here the surface in contact is constantly changing. Frictional force (F) is proportional to the force holding the bodies together (the 'normal reaction' R). The constants of proportionality (for different cases) are called *coefficients of friction* (symbol: μ):

$$\mu = F/R$$

Two *laws of friction* are sometimes stated:
(1) The frictional force is independent of the area of contact (for the same force holding the surfaces together).
(2) The frictional force is proportional to the force holding the surfaces together. In sliding friction it is independent of the relative velocities of the surfaces.

frustrum A geometric solid produced by two parallel planes cutting the solid, or by one plane parallel to the base.

fulcrum The point about which a lever turns.

function (mapping) Any defined procedure that relates one number, quantity, etc., to another or others. In algebra, a function of a variable x is often written as $f(x)$. If two variable quantities, x and y, are related by the equation $y = x^2 + 2$, for example, then y is a function of x or $y = f(x) = x^2 + 2$. The function here means 'square the number and add two'. x is the *independent variable* and y is the *dependent variable*. The *inverse function* – the one that expresses x in terms of y in this case – would be $x = \pm\sqrt{(y - 2)}$, which might be written as $x = g(y)$.
A function can be regarded as a relationship between the elements of one set (the *range*) and those of another set (the *domain*). For each element of the first set there is a corresponding element of the second set into which it is 'mapped' by the function. For example, the set of numbers {1,2,3,4} is mapped into the set {1,8,27,64} by taking the cube of each element. A function may also map elements of a set into others in the same set. Within

the set {all women}, there are two subs {mothers} and {daughters}. The ma ping between them is 'is the mother of and the inverse is 'is the daughter of'.

fundamental The simplest way (mode) in which an object can vibrate. The fundamental frequency is the frequency of this vibration. The less simple modes of vibration are the higher *harmonics*; their frequencies are higher than that of the fundamental.

fundamental theorem of algebra
Every polynomial equation of the form:
$$a_0 z^n + a_1 z^{n-1} + a_2 z^{n-2} + \dots$$
$$a_{n-1} z + a_n = 0$$
in which a_0, a_1, a_2, etc., are complex numbers, has at least one complex root. *See also* polynomial.

fundamental theorem of calculus
The theorem used in calculating the value of a definite integral. If $f(x)$ is a continuous function of x in the interval $a \leqslant x \leqslant b$, and if $g(x)$ is any indefinite integral of $f(x)$, then:

$$\int_a^b f(x)dx = [g(x)|_a^b] = g(a) - g(b)$$

See also integral, definite integral, indefinite integral.

fundamental units The units of length, mass, and time that form the basis of most systems of units. In SI, the fundamental units are the metre, the kilogram, and the second. *See also* base unit.

furlong A unit of length equal to one eighth of a mile. It is equivalent to 201.168 m.

G

gallon A unit of capacity usually used to measure volumes of liquids. In the UK it is defined as the space occupied by 10 pounds of pure water and is equal to $4.5461 \times 10^{-3} \, \mathrm{m}^3$. In the US it is defined as 231 cubic inches and is equal to $3.7854 \times 10^{-3} \, \mathrm{m}^3$. 1 UK gallon is equal to 1.2 US gallons.

game theory A mathematical theory of the optimal behaviour in competitive situations in which the outcomes depend not only on the participants' choices but also on chance and the choices of others. A *game* may be defined as a set of rules describing a competitive situation involving a number of competing individuals or groups of individuals. These rules give their permissible actions at each stage of the game, the amount of information available, the probabilities associated with the chance events that might occur, the circumstances under which the competition ends, and a pay-off scheme specifying the amount each player pays or receives at such a conclusion. It is assumed that the players are rational in the sense that they prefer better rather than worse outcomes and are able to place the possible outcomes in order of merit. Game theory has applications in military science, economics, politics, and many other fields.

gamma function The integral function

$$\Gamma(x) = \int_0^\infty t^{x-1} e^{-t} \mathrm{d}t$$

If x is a positive integer n, then $\Gamma(n) = n!$ If x is an integral multiple of $\frac{1}{2}$, the function is a multiple of $\sqrt{\pi}$.

$$\Gamma(\tfrac{1}{2}) = \sqrt{\pi}$$
$$\Gamma(3/2) = (\tfrac{1}{2})\sqrt{\pi}$$
etc.

gauss Symbol: G The unit of magnetic flux density in the c.g.s. system. It is equal to 10^{-4} tesla.

Gaussian distribution *See* normal distribution.

general conic *See* conic.

general form (of an equation) A formula that defines a type of relationship between variables but does not specify values for constants. For example, the general form of a polynomial equation in x is

$$ax^n + bx^{n-1} + cx^{n-2} + \ldots = 0$$

a, b, c, etc., are constants and n is the highest integer power of x, called the *degree* of the polynomial. Similarly, the general form of a quadratic equation is

$$ax^2 + bx + c = 0$$

See also equation, polynomial.

general theory *See* relativity, theory of.

generator A line that generates a surface; for example, in a cone, cylinder, or solid of revolution.

geodesic A line on a surface between two points that is the shortest distance between the points. On a plane a geodesic is a straight line. On a spherical surface it is part of a great circle of the sphere.

geometric distribution The distribution of the number of independent Bernoulli trials before a successful result is obtained; for example, the distribution of the number of times a coin has to be tossed before a head comes up. The probability that the number of trials (x) is k is

$$P(x = k) = q^{k-1} p$$

The mean and variance are $1/p$ and q/p^2 respectively. The moment generating function is $e^t p / (1 - q e^t)$.

geometric mean *See* mean.

geometric progression *See* geometric sequence.

geometric sequence (geometric progression) A sequence in which the ratio of each term to the one after it is constant, for example, 1, 3, 9, 27, ... The general formula for the nth term of a geometric sequence is $u_n = ar^n$. The ratio is called the *common ratio*. In the example the first term, a, is 1, the common ratio, r, is 3, and so u_n equals 3^n. If a geometric sequence is convergent, r lies between 1 and -1 (exclusive) and the limit of the sequence is 0. That is, u_n approaches zero as n becomes infinitely large. *Compare* arithmetic sequence. *See also* geometric series, convergent sequence, divergent sequence, sequence.

geometric series A series in which the ratio of each term to the one after it is constant, for example, $1 + 2 + 4 + 8 + 16 + \ldots$. The general formula for a geometric series is

$$S_n = a + ar + ar^2 + \ldots + ar^n = a(r^n - 1)/(r - 1)$$

In the example, the first term, a, is 1, the *common ratio*, r, is 2, and so the nth term ar^n equals 2^n. If r is greater than 1, the series will not be convergent. If $-1 < r < 1$ and the sum of all the terms after the nth term can be made as small as required by making n large enough, then the series is convergent. This means that there is a finite sum even when n is infinitely large. The *sum to infinity* of a convergent geometric series is $a/(1 - r)$. *Compare* arithmetic series. *See also* geometric sequence, convergent series, divergent series, series.

geometry The branch of mathematics concerned with points, lines, curves, and surfaces – their measurement, relationships, and properties that are invariant under a given group of transformations. For example, geometry deals with the measurement or calculation of angles between straight lines, the basic properties of circles, and the relationship between lines and points on a surface. *See* Euclidean geometry, non-Euclidean geometry, analytical geometry, topology.

giga- Symbol: G A prefix denoting 10^9. For example, 1 gigahertz (GHz) = 10^9 hertz (Hz).

gill A unit of capacity equal to one quarter of a pint. A UK gill is equivalent to 1.420 × 10^{-4} m³ and a US gill is equivalent to 1.182 9 × 10^{-4} m³. *See* pint.

Gödel's incompleteness theorem A fundamental result of mathematical logic showing that any formal system powerful enough to express the truths of arithmetic must be incomplete; that is that it will contain statements that are *true* but cannot be proved using the system itself.

golden rectangle A rectangle in which the adjacent sides are in the ratio $(1 + \sqrt{5})/2$.

golden section The division of a line of length l into two lengths a and b so that $l/a = a/b$, that is, $a/b = (1 + \sqrt{5})/2$. Proportions based on the golden section are particularly pleasing to the eye and occur in many paintings, buildings, designs, etc.

Goldbach conjecture The conjecture that every even number other than 2 is the sum of two prime numbers; so far, unproved.

governor A mechanical device to control the speed of a machine. One type of simple governor consists of two loads attached to a shaft so that as the speed of rotation of the shaft increases, the loads move further outwards from the centre of rotation, while still remaining attached to the shaft. As they move outwards they operate a control that reduces the rate of fuel or energy input to the machine. As they reduce speed and move inwards they increase the fuel or energy input. Thus, on the principle of negative feedback, the speed of the machine is kept fairly constant under varying conditions of load.

grad (gradient) Symbol: ∇ A vector operator that, for any function $f(x,y,z)$, has components in the x, y, and z directions

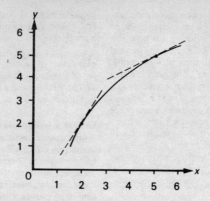

The gradient of the curve at the point (2,2) is 2, and at the point (5,5) is 1/2.

equal to the partial derivatives with respect to x, y, and z in that order. It is defined as:

$$\text{grad } f = \nabla f = \mathbf{i}\partial f/\partial x + \mathbf{j}\partial f/\partial y + \mathbf{k}\partial f/\partial z$$

where $\mathbf{i}$, $\mathbf{j}$, and $\mathbf{k}$ are the unit vectors in the x, y, and z directions. In physics, ∇F is often used to describe the spatial variation in the magnitude of a force F in, for example, a magnetic or gravitational field. It is a vector with the direction in which the rate of change of F is a maximum, if such a maximum exists. In the Earth's gravitational field this would be radially towards the centre of the Earth (downwards). In a magnetic field, ∇F would point along the lines of force. See also partial derivative.

grade Symbol: g A unit of plane angle equal to one hundredth of a right angle. 1^g is equal to $0.9°$.

gradient 1. (slope) In rectangular Cartesian coordinates, the rate at which the y-coordinate of a curve or a straight line changes with respect to the x-coordinate. The straight line $y = 2x + 4$ has a gradient of $+2$; y increases by two for every unit increase in x. The general equation of a straight line is $y = mx + c$, where m is the gradient and c is a constant ($(0,c)$ is the point at which the line cuts the y-axis, i.e.

the intercept). If m is negative, y decreases as x increases.

For a curve, the gradient changes continuously; the gradient at a point is the gradient of the straight line that is a tangent to the curve at that point. For the curve $y = f(x)$, the gradient is the derivative dy/dx. For example, the curve $y = x^2$ has a gradient given by $dy/dx = 2x$, at any particular value of x. See also derivative.

2. See grad.

gram Symbol: g A unit of mass defined as 10^{-3} kilogram.

gram-atom See mole.

gramme An alternative spelling of gram.

gram-molecule See mole.

graph 1. A drawing that shows the relationship between numbers or quantities. Graphs are usually drawn with coordinate axes at right angles. For example, the heights of children of different ages can be shown by making the distance along a horizontal line represent the age in years and the distance up a vertical line represent the height in metres. A point marked on the graph ten units along and 1.5 units up represents a ten-year-old who is 1.5 metres tall. Similarly, graphs are used to give a geometric representation of equations. The graph of $y = x^2$ is a parabola. The graph of $y = 3x + 10$ is a straight line. Simultaneous equations can be solved by drawing the graphs of the equations, and finding the points where they cross. For the two equations above, the graphs cross at two points: $x = -2$, $y = 4$ and $x = 5$, $y = 25$.

There are various types of graph. Some, such as the histogram and the pie chart, are used to display numerical information in a form that is simple and quickly understood. Some, such as conversion graphs, are used as part of a calculation. Others, such as scatter diagrams, may be used in analysing the results of a scientific experiment. *See also* bar chart, conversion graph, histogram, pie chart, scatter diagram.

2. (topology) A network of lines and vertices. *See* Königsberg bridge problem.

graphics display (graphical display unit) *See* visual display unit.

gravitation The concept originated by Isaac Newton around 1666 to account for the apparent motion of the Moon round the Earth, the essence being a force of attraction, called gravity, between the Moon and the Earth. Newton used this theory of gravitation to give the first satisfactory explanations of many facts, such as Kepler's laws c f planetary motion, the ocean tides, and the precession of the equinoxes. *See also* Newton's law of universal gravitation.

gravitational constant Symbol: G The constant of proportionality in the equation that expresses Newton's law of universal gravitation:

$$F = Gm_1m_2/r^2$$

where F is the gravitational attraction between two point masses m_1 and m_2 separated by a distance r. The value of G is 6. 67×10^{-11} N m^2 kg^{-2}. It is regarded as a universal constant, although it has been suggested that the value of G may be changing slowly owing to the expansion of the Universe. *See also* Newton's law of universal gravitation.

gravitational field The region of space in which one body attracts other bodies as a result of their mass. To escape from this field a body has to be projected outwards with a certain speed (the *escape speed*). The strength of the gravitational field at a point is given by the ratio force/ mass, which is equivalent to the acceleration of free fall, g. This may be defined as GM/r^2, where G is the gravitational constant, M the mass of the object at the centre of the field, and r the distance between the object and the point in question. The standard value of the acceleration of free fall at the Earth's surface is 9.8 m s^{-2}, but it varies with altitude (i.e. with r^2).

gravitational mass The mass of a body as measured by the force of attraction between masses. The value is given by Newton's law of universal gravitation. Inertial and gravitational masses appear to be equal in a uniform gravitational field. *See also* inertial mass, mass.

gravity The gravitational pull of the Earth (or other celestial body) on an object. The force of gravity on an object causes its weight.

gravity, centre of *See* centre of mass.

great circle A circle on the surface of a sphere that has the same radius as the sphere. A great circle is formed by a cross-section by any plane that passes through the centre of the sphere.

greatest upper bound *See* bound.

gross 1. Denoting a weight of goods including the weight of the container or packing.
2. Denoting a profit calculated before deducting overhead costs, expenses, and (usually) taxes.
Compare net.

group A set having certain additional properties:

group speed

(1) In a group there is a binary operation for which the elements of the set can be related in pairs, giving results that are also members of the group (the property of *closure*). For example, the set of all positive and negative numbers and zero form a group under the operation of addition. Adding any member to any other gives an element that is also a member of the group; e.g. $3 + (-2) = 1$, etc.

(2) There is an identity element for the operation – i.e. an element that, combined with another, leaves it unchanged. In the example, the identity element is zero: adding zero to any member leaves it unchanged; $3 + 0 = 3$, etc.

(3) For each element of the group there is another element – its *inverse*. Combining an element with its inverse leads to the identity element. In the example, the number $+3$ has an inverse -3 (and vice versa); thus $+3 + (-3) = 0$.

(4) The associative law holds for the members of the group. In the example:

$$2 + (3 + 5) = (2 + 3) + 5$$

Any set of elements obeying the above rules forms a group. Note that the binary operation need not be addition. *Group theory* is important in many branches of mathematics – for instance, in the theory of roots of equations. It is also very useful in diverse branches of science. In chemistry, group theory is used in describing symmetries of molecules to determine their energy levels and explain their spectra. In physics, certain elementary particles can be classified into mathematical groups on the basis of their quantum numbers (this led to the discovery of the omega-minus particle as a missing member of a group). Group theory has also been applied to linguistics.
See also Abelian group, cyclic group.

group speed If a wave motion has a phase speed that depends on wavelength, the disturbance of a progressive wave travels with a different speed than the phase speed. This is called the *group speed*. It is the speed with which the group of waves travels, and is given by:

$$U = c - \lambda dc/d\lambda$$

where c is the phase speed. The group speed is the one that is usually obtained by measurement. If there is no dispersion of the wave motion, as for electromagnetic radiation in free space, the group and phase speeds are equal.

gyroscope A rotating object that tends to maintain a fixed orientation in space. For example, the axis of the rotating Earth always points in the same direction towards the Pole Star (except for a small precession). A spinning top or a cyclist are stable when moving at speed because of the gyroscopic effect. Practical applications are the navigational gyrocompass and automatic stabilizers in ships and aircraft. *See also* precession.

H

half-angle formulae *See* addition formulae.

Hamiltonian In classical mechanics, a function of the coordinates q_i, $i = 1, 2, \ldots n$, and momenta, p_i, $i = 1, 2, \ldots n$, generally denoted by H and defined by

$$H = \Sigma p_i q_i - L$$

where q_i denotes the derivative with respect to time and L is the *Lagrangian function* of the system expressed as a function of the coordinates momenta and time. If the Lagrangian function does not depend explicitly on time, the system is said to be *conservative* and H is the total energy of the system.

ham-sandwich theorem 1. A ham sandwich can be cut with one stroke of a knife so that the ham and each slice of bread are exactly cut in half. More formally, if A, B, and C are bounded connected sets in space, then there is a plane that cuts each set into two sets with equal volume.
2. If $f(x) \leqslant g(x) \leqslant h(x)$ for all x and the functions f and h have the same limit then g also has this limit.

hardware The physical embodiment of a computer system, i.e. its electronic circuitry, disk and magnetic tape units, line printers, cabinets, etc. *Compare* software.

harmonic mean *See* mean.

harmonic motion A regularly repeated sequence that can be expressed as the sum of a set of sine waves. Each component sine wave represents a possible simple harmonic motion. The complex vibration of sound sources (with fundamental and overtones), for instance, is a harmonic motion, as is the sound wave produced. *See also* simple harmonic motion.

harmonic analysis The use of trigonometric series to study mathematical functions. *See* Fourier series.

harmonic progression *See* harmonic sequence.

harmonic sequence (harmonic progression) An ordered set of numbers, the reciprocals of which have a constant difference between them; for example, $\{1, \frac{1}{2}, \frac{1}{3}, \frac{1}{4}, \ldots 1/n\}$. In this example $\{1, 2, 3, 4, \ldots n\}$ have a constant difference — i.e. they form an arithmetic sequence. The reciprocals of the terms in a harmonic sequence form an arithmetic sequence, and vice versa. *See also* arithmetic sequence.

harmonic series The sum of the terms in a harmonic sequence; for example: $1 + \frac{1}{2} + \frac{1}{3} + \frac{1}{4} + \ldots$

HCF Highest common factor. *See* common factor.

hecto- Symbol: h A prefix denoting 10^2. For example, 1 hectometre (hm) = 10^2 metres (m).

height A vertical distance, usually upwards, from a base line or plane. For example, the perpendicular distance from the base of a triangle to the vertex opposite, and the distance between the uppermost and the base planes of a cuboid, are both known as the height of the figure.

helix A spiral-shaped space curve. A *cylindrical helix* lies on a cylinder. A *conical helix* lies on a cone. For example, the shape of the thread on a screw is a helix. In a straight screw, it is a cylindrical helix and in a conically tapered screw it is a conical helix.

hemisphere The surface bounded by half of a sphere and a plane through the centre of the sphere. *See* sphere.

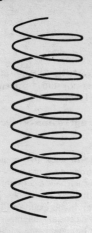

Cylindrical
helix

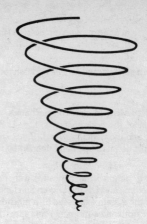

Conical
helix

henry Symbol: H The SI unit of inductance, equal to the inductance of a closed circuit that has a magnetic flux of one weber per ampere of a current in the circuit. 1 $H = 1 \, Wb \, A^{-1}$.

heptagon A plane figure with seven straight sides. A *regular heptagon* has seven equal sides and seven equal angles.

Hermitian matrix The *Hermitian conjugate* of a matrix is the transpose of the complex conjugate of the matrix, where the complex conjugate of a matrix is the matrix whose elements are the complex conjugates of the corresponding elements of the given matrix (*see* conjugate complex numbers). A *Hermitian matrix* is a matrix that is its own Hermitian conjugate; i.e. a square matrix such that a_{ij} is the complex conjugate of a_{ij} for all i and j where a_{ij} is the element in the ith row and jth column.

Hero's formula A formula for the area of a triangle with sides a, b, and c:
$$A = \sqrt{[s(s - a)(s - b)(s - c)]}$$
where s is half the perimeter; i.e. $\frac{1}{2}(a + b + c)$.

hertz Symbol: Hz The SI unit of frequency, defined as one cycle per second (s^{-1}). Note that the hertz is used for regularly repeated processes, such as vibration or wave motion. An irregular process, such as radioactive decay, would have units expressed as s^{-1} (per second).

heuristic Based on trial and error, as for example some techniques in iterative calculations. *See also* iteration.

hexadecimal Denoting or based on the number sixteen. A hexadecimal number is made up with sixteen different digits instead of the ten in the decimal system. Normally these are 0, 1, 2, 3, 4, 5, 6, 7, 8, 9, A, B, C, D, E, F. For example, 16 is written as 10, 21 is written as 15 (16 + 5), 59 is written as 3B [(3 × 16) + 11]. Hexadecimal numbers are sometimes used in computer systems, because they are much shorter than the long strings of binary digits that the machine normally uses. Binary numbers are easily converted into hexadecimal numbers by grouping the digits in fours. *Compare* binary, decimal, duodecimal, octal.

hexagon A plane figure with six straight sides. A *regular hexagon* is one with all six sides and all six angles equal, the angles all being 120°. Congruent regular hexagons can be fitted together to cover completely a plane surface. Apart from squares and equilateral triangles, they are the only regular polygons with this property.

hexahedron A polyhedron that has six faces. For example, the cube, the cuboid, and the rhombohedron are all hexahedrons. The cube is a *regular hexahedron*; all six faces are congruent squares. *See also* polyhedron.

highest common factor *See* common factor.

high-level language *See* program.

histogram A statistical graph that represents, by the height of a rectangular column, the number of times that each class of result occurs in a sample or experiment. *See also* frequency polygon.

holomorphic *See* analytic.

homeomorphism A one-one transformation between two topological spaces that is continuous in both directions. What this means is that if two figures are homeomorphic one can be continuously deformed into the other without tearing. For example, any two spheres of any size are homeomorphic. But a sphere and a torus are not.

homogeneous 1. (in a function) Having all the terms to the same degree in the variables. For a homogeneous function $f(x,y,z, \ldots)$ of degree n
$$F(kx,ky,kz, \ldots) = k^n f(x,y,z)$$
for all values of k. For example, $x^2 + xy + y^2$ is a homogeneous function of degree 2 and
$$(kx)^2 + kx.ky + (ky)^2 = k^2(x^2 + xy + y^2)$$
2. Describing a substance or object in which the properties do not vary with posi-

tion; in particular, the density is constant throughout.

homomorphism If S and T are sets on which binary relations • and · are defined respectively, a mapping h from S and T is a *homomorphism* if it satisfies the condition $h(x•y) = h(x)·h(y)$ for all x and y in S, i.e. it preserves structure. If the mapping is one-to-one it is called an *isomorphism* and the sets S and T are *isomorphic*.

horse power Symbol: HP A unit of power equal to 550 foot-pounds per second. It is equivalent to 746 W.

hundredweight Symbol: cwt In Britain, a unit of mass equal to 112 pounds. It is equivalent to 50.802 3 kg. In the USA a hundredweight is equal to 100 pounds, but this unit is rarely used.

hybrid computer A computer system containing both analog and digital devices so that the properties or each can be used to the greatest advantage. For instance, a digital and an analog computer can be interconnected so that data can be transferred between them. This is achieved by means of a *hybrid interface*. Hybrid computers are designed for specific tasks and have a variety of uses, mainly in scientific and technical fields. *See also* computer, analog computer.

hydraulic press A machine in which forces are transferred by way of pressure in a fluid. In a hydraulic press the effort F_1 is applied over a small area A_1 and the load F_2 exerted over a larger area A_2. Since the pressure is the same, $F_1/A_1 = F_2/A_2$. The force ratio for the machine, F_2/F_1, is A_1/A_2. Thus, in this case (and in the related hydraulic braking system and hydraulic jack) the force exerted by the user is less than the force applied; the force ratio is greater than 1. If the distance moved by the effort is s_1 and that moved by the load is s_2 then, since the same volume is transmitted through the system, $s_1 A_1 = s_2 A_2$; i.e. the distance ratio is A_2/A_1. In practice, the device is not very efficient since frictional effects are large. *See* machine.

hydrostatics The study of fluids (liquids and gases) in equilibrium.

hyperbola A conic with an eccentricity greater than 1. The hyperbola has two branches and two axes of symmetry. An axis through the foci cuts the hyperbola at two *vertices*. The line segment joining these vertices is the *transverse axis* of the hyperbola. The *conjugate axis* is a line at right angles to the transverse axis through the centre of the hyperbola. A chord through a focus perpendicular to the transverse axis is a *latus rectum*.
In Cartesian coordinates the equation:
$$x^2/a^2 - y^2/b^2 = 1$$
represents a hyperbola with its centre at the origin and the transverse axis along the x-axis. $2a$ is the length of the transverse axis. $2b$ is the length of the conjugate axis. This is the distance between the vertices of a different hyperbola (the *conjugate hyperbola*) with the same asymptotes as the given one. The foci of the hyperbola are at the points (ae,o) and $(-ae,o)$, where e is the eccentricity. The asymptotes have the equations:
$$x/a - y/b = 0$$
$$x/a + y/b = 0$$
The equation of the conjugate hyperbola is
$$x^2/a^2 - y^2/b^2 = -1$$
The length of the latus rectum is $2b^2/ae$. A hyperbola for which a and b are equal is a rectangular hyperbola:
$$x^2 - y^2 = a^2$$
If a rectangular hyperbola is rotated so that the x- and y-axes are asymptotes, then its equation is
$$xy = k$$
where k is a constant.
See also conic.

hyperbolic functions A set of functions that have properties similar in some ways to the trigonometric functions, called the hyperbolic sine, hyperbolic cosine, etc. They are related to the hyperbola in the way that the trigonometric functions (circular functions) are related to the circle.

The *hyperbolic sine* (sinh) of an angle a is defined as:
$$\sinh a = \tfrac{1}{2}(e^a - e^{-a})$$
The *hyperbolic cosine* (cosh) of an angle a is defined as:
$$\cosh a = \tfrac{1}{2}(e^a + e^{-a})$$
The *hyperbolic tangent* (tanh) of an angle a is defined as:
$$\tanh a = \sinh a/\cosh a =$$
$$(e^a - e^{-a})/(e^a + e^{-a})$$
Hyperbolic secant (sech), *hyperbolic cosecant* (cosech), and *hyperbolic cotangent* (coth) are defined as the reciprocals of cosh, sinh, and tanh respectively. Some of the fundamental relationships between hyperbolic functions are:
$$\sinh(-a) = -\sinh a$$
$$\cosh(-a) = +\cosh a$$
$$\cosh^2 a - \sin^2 a = 1$$
$$\operatorname{sech}^2 a + \tanh^2 a = 1$$
$$\coth^2 a - \operatorname{cosech}^2 a = 1$$

hyperboloid A surface generated by rotating a hyperbola about one of its axes of symmetry. Rotation about the conjugate axis gives a *hyperboloid of one sheet*. Rotation about the transverse axis gives a *hyperboloid of two sheets*.

hypotenuse The side opposite the right angle in a right-angled triangle. The ratios of the hypotenuse length to the lengths of the other sides are used in trigonometry to define the sine and cosine functions of angles.

hypothesis A statement, theory, or formula that has yet to be proved but is assumed to be true for the purposes of the argument.

hypothesis test (significance test) A rule for deciding whether an assumption (hypothesis) about the distribution of a random variable should be accepted or rejected, using a sample from the distribution. The assumption is called the null hypothesis, written H_0, and it is tested against some alternative hypothesis, H_1. For example, when a coin is tossed H_0 can be P(heads) = $\tfrac{1}{2}$ and H_1 that P(heads) > $\tfrac{1}{2}$. A statistic is computed from the sample data. If it falls in the critical region, where the value of the statistic is significantly dif-

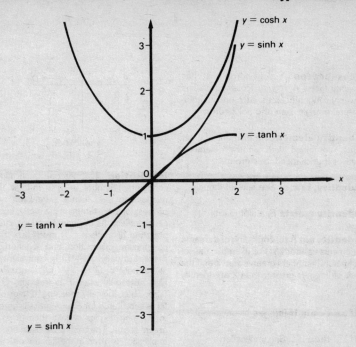

The graphs of the hyperbolic functions cosh x, sinh x, and tanh x.

ferent from that expected under H_0, H_0 is rejected in favour of H_1. Otherwise H_0 is accepted. A type I error occurs if H_0 is rejected when it should be accepted. A type II error occurs if it is accepted when it should be rejected. The significance level of the test, α, is the maximum probability with which a type I error can be risked. For example, $\alpha = 1\%$ means H_0 is wrongly rejected in one case out of 100.

I

icosahedron A polyhedron that has twenty faces. A *regular icosahedron* has twenty congruent faces, each one an equilateral triangle. *See also* polyhedron.

identity element An element of a set that, combined with another element, leaves it unchanged. *See* group.

identity, law of *See* laws of thought.

identity matrix *See* unit matrix.

identity set A set consisting of the same elements as another. For example, the set of natural numbers greater than 2 and the set of integers greater than 2 are identity sets.

if and only if (iff) *See* biconditional.

if . . . then . . . *See* implication.

image The result of a geometrical transformation or a mapping. For example, in geometry, when a set of points are transformed into another set by reflection in a line, the reflected figure is called the image. Similarly, the result of a rotation or a projection is called the image. The algebraic equivalent of this occurs when a function f(x) acts on a set A of values of x to produce an image set B. *See also* domain, range, transformation.

imaginary number A multiple of i, the square root of minus one. The use of imaginary numbers is needed to solve equations such as $x^2 + 2 = 0$, for which the solutions are $x = +i\sqrt{2}$ and $x = -i\sqrt{2}$. *See* complex number.

Imperial units The system of measurement based on the yard and the pound, formerly used in the UK. The f.p.s. system was a scientific system based on Imperial units.

P	Q	$P \rightarrow Q$
T	T	T
T	F	F
F	T	T
F	F	T

implication

implication 1. (material implication, conditional) Symbol: $\rightarrow$ or $\supset$ In logic, the relationship *if . . . then . . .* between two propositions or statements. Strictly, implication reflects its ordinary language interpretation (*if . . . then . . .*) much less than conjunction, disjunction, and negation do theirs. Formally, $P \rightarrow Q$ is equivalent to 'either not P or Q' ($\sim P \vee Q$), hence $P \rightarrow Q$ is false *only* when P is true and Q is false. Thus, logically speaking, if 'pigs can fly' is substituted for P and 'grass is green' for Q then 'if pigs can fly then grass is green' is true. The truth table definition of implication is given in the illustration.
2. In algebra, the symbol $\Rightarrow$ is used between two equations when the first implies the second. For example:
$$x = y \Rightarrow x^2 = y^2$$
See also condition, truth table.

implicit Denoting a function that contains two or more variables that are not independent of each other. An *implicit function* of x and y is one of the form f(x,y) = 0, for example,
$$x^2 + y^2 - 4 = 0$$
Sometimes an *explicit function*, that is, one expressed in terms of an independent variable, can be derived from an implicit function. For example,
$$y + x^2 - 1 = 0$$
can be written as
$$y = 1 - x^2$$
where y is an explicit function of x.

improper fraction *See* fraction.

Inscribed circle

impulse (impulsive force) A force acting for a very short time, as in a collision. If the force (F) is constant the impulse is $F\delta t$, δt being the time period. If the force is variable the impulse is the integral of this over the short time period. An impulse is equal to the change of momentum that it produces.

impulsive force See impulse.

incentre See incircle.

inch Symbol: in or '' A unit of length equal to one twelfth of a foot. It is equivalent to 0.025 4 m.

incircle (inscribed circle) A circle drawn inside a figure, touching all the sides. The centre of the circle is the *incentre* of the figure. *Compare* circumcircle.

inclined plane A type of machine. Effectively a plane at an angle, it can be used to raise a weight vertically by movement up an incline. Both distance ratio and force ratio depend on the angle of inclination (θ) and equal $1/\sin\theta$. The efficiency can be fairly high if friction is low. The screw and the wedge are both examples of inclined planes. *See* machine.

inclusion See subset.

inclusive disjunction (inclusive or) See disjunction.

increment A small difference in a variable. For example, x might change by an increment Δx from the value x_1 to the value x_2; $\Delta x = x_2 - x_1$. In calculus, infinitely small increments are used. *See also* differentiation, integration.

indefinite integral The general integration of a function f(x), of a single variable, x, without specifying the interval of x to which it applies. For example, if f(x) = x^2, the indefinite integral
$$\int f(x)dx = \int x^2 dx = x^3/3 + C$$
where C is an unknown constant (the constant of integration) that depends on the interval. *Compare* definite integral. *See also* integration.

independence See probability.

independent See dependent.

independent variable See variable.

indeterminate equation An equation that has an infinite number of solutions. For example,

93

$$x + 2y = 3$$

is indeterminate because an infinite number of values of x and y will satisfy it. An indeterminate equation in which the variables can take only integer values is called a *Diophantine equation* and it has an infinite but denumerable set of solutions.

indeterminate form An expression that can have no quantitative meaning; for example $0/0$.

index A number that indicates a characteristic or function in a mathematical expression. For example, in y^4, the exponent, 4, is also known as the index. Similarly in $\sqrt[3]{27}$ and $\log_{10}x$, the numbers 3 and 10 respectively are called indices (or indexes).

indirect proof (reductio ad absurdum) A logical argument in which a proposition or statement is proved by showing that its negation or denial leads to a contradiction. *Compare* direct proof. *See* contradiction.

induction 1. (mathematical induction) A method of proving mathematical theorems, used particularly for series sums. For instance, it is possible to show that the series $1 + 2 + 3 + 4 + \ldots$ has a sum to n terms of $n(n + 1)/2$. First we show that if it is true for n terms it must also be true for $(n + 1)$ terms. According to the formula

$$S_n = n(n + 1)/2$$

if the formula is correct, the sum to $(n + 1)$ terms is obtained by adding $(n + 1)$ to this

$$S_{n+1} = n(n + 1)/2 + (n + 1)$$
$$S_{n+1} = (n + 1)(n + 2)/2$$

This agrees with the result obtained by replacing n in the general formula by $(n + 1)$, i.e.:

$$S_{n+1} = (n + 1)(n + 1 + 1)/2$$
$$S_{n+1} = (n + 1)(n + 2)/2$$

Thus, the formula is true for $(n + 1)$ terms if it is true for n terms. Therefore, if it is true for the sum to one term $(n = 1)$, it must be true for the sum to two terms $(n + 1)$. Similarly, if true for two terms, it

must be true for three terms, and so on through all values of n. It is easy to show that it is true for one term:

$$S_n = 1(1 + 1)/2$$
$$S_n = 1$$

which is the first term in the series. Hence the theorem is true for all integer values of n.

2. In logic, a form of reasoning from individual cases to general ones, or from observed instances to unobserved ones. Inductive arguments can be of the form: F_1 is A, F_2 is A ... F_n is A, therefore all Fs are A ('this swan has wings, that swan has wings ... therefore all swans have wings'); or: all Fs observed so far are A, therefore all Fs are A ('all swans observed so far are white, therefore all swans are white'). Unlike deduction, asserting the premises while denying the conclusion in an induction does not lead to a contradiction. The conclusion is *not* guaranteed to be true if the premises are. *Compare* deduction. *See* contradiction.

inelastic collision A collision for which the restitution coefficient is less than one. In effect, the relative velocity after the collision is less than that before; the kinetic energy of the bodies is not conserved in the collision, even though the system may be closed. Some of the kinetic energy is converted into internal energy. *See also* restitution, coefficient of.

inequality A relationship between two expressions that are not equal, often written in the form of an equation but with the symbols $>$ or $<$ meaning 'is greater than' and 'is less than'. For example, if $x < 4$ then $x^2 < 16$. If $y^2 > 25$, then $y > 5$ or $y < -5$. If the end values are included, the symbols $\geq$ (is greater than or equal to) and $\leq$ (is less than or equal to) are used. When one quantity is very much smaller than another, it is shown by $<<$ or $>>$. For example, if x is a large number $x >> 1/x$ or $1/x << x$. *See also* equality.

inertia An inherent property of matter implied by Newton's first law of motion: inertia is the tendency of a body to resist change in its motion. *See also* inertial mass, Newton's laws of motion.

inertial mass The mass of an object as measured by the property of inertia. It is equal to the ratio force/acceleration when the object is accelerated by a constant force. In a uniform gravitational field, it appears to be equal to gravitational mass – all objects have the same gravitational acceleration at the same place. *See also* gravitational mass.

inertial system A frame of reference in which an observer sees an object that is free of all external forces to be moving at constant velocity. The observer is called an *inertial observer*. Any frame of reference that moves with constant velocity and without rotation relative to an inertial frame is also an inertial frame. Newton's laws of motion are valid in any inertial frame (but not in an accelerated frame), and the laws are therefore independent of the velocity of an inertial observer. *See also* frame of reference, Newton's laws of motion.

inf *See* infimum.

inference 1. The process of reaching a conclusion from a set of premises in a logical argument. An inference may be deductive or inductive. *See also* deduction, induction.
2. *See* sampling.

infimum (inf) The greatest lower bound of a set. *See* bound.

infinite number The smallest infinite number is $\aleph_0$ (aleph zero). This is the number of members in the set of integers. A whole hierarchy of increasingly large infinite numbers can be defined on this basis. $\aleph_1$, the next largest, is the number of subsets of the set of integers. *See also* continuum, aleph, countable.

infinite sequence *See* sequence.

infinite series *See* series.

infinite set A set in which the number of elements is infinite. For example, the set of 'positive integers', $z = \{1, 2, 3, 4, \ldots\}$, is infinite but the set of 'positive integers less than 20' is a *finite set*. Another example of an infinite set is the number of circles in a particular plane. *Compare* finite set.

infinitesimal Infinitely small, but not equal to zero. Infinitesimal changes or differences are made use of in calculus (infinitesimal calculus). *See* calculus.

infinity Symbol: ∞ The value of a quantity that increases without limit. For example, if $y = 1/x$, then y becomes infinitely large, or approaches infinity, as x approaches 0. An infinitely large negative quantity is denoted by $-\infty$ and an infinitely large positive quantity by $+\infty$. If x is positive, $(y = -(1/x))$ tends to $-\infty$ as x tends to 0.

inflection *See* point of inflection.

information theory The branch of probability theory that deals with uncertainty, accuracy, and information content in the transmission of messages. It can be applied to any system of communication, including electrical signals and human speech. Random signals (noise) are often added to a message during the transmission process, altering the signal received from that sent. Information theory is used to work out the probability that a particular signal received is the same as the signal sent. Redundancy, for example simply repeating a message, is needed to overcome the limitations of the system. Redundancy can also take the form of a more complex checking process. In transmitting a sequence of numbers, their sum might also be transmitted so that the receiver will know that there is an error when the sum does not correspond to the rest of the message. The sum itself gives no extra information since, if the other numbers are correctly received, the sum can easily be calculated. The statistics of choosing a message out of all possible messages (letters in the alphabet or binary digits for example) determines the amount of information contained in it. Information is measured in bits (binary digits). If one out of two possible signals are sent then the information content is one bit. A choice of one out of four possible signals contains more information, although the signal itself might be the same.

inner product Consider a vector space V over a scalar field F. An *inner product* on V is a mapping of ordered pairs of vectors in V into F; i.e. with every pair of vectors x and y there is associated a scalar, which is written $\langle x,y \rangle$ and called the inner product of x and y, such that for all vectors x, y, z and scalars α

(i) $\langle x+y,z \rangle = \langle x,z \rangle + \langle y,z \rangle$

(ii) $\langle x,y \rangle = \alpha \langle x,y \rangle$

(iii) $\langle x,y \rangle = \overline{\langle y,x \rangle}$, where $\overline{\langle a,b \rangle}$ is the complex conjugate of $\langle a,b \rangle$.

(iv) $\langle x,x \rangle \geqslant 0$, $\langle x,x \rangle = 0$ if and only if $x = 0$.

An inner product on V defines a *norm* on V given by $\|x\| = \sqrt{\langle x,x \rangle}$. *See* norm.

input 1. The signal or other form of information that is applied (fed in) to an electrical device, machine, etc. The input to a computer is the data and programmed instructions that a user communicates to the machine. An *input device* accepts computer input in some appropriate form and converts the information into a code of electrical pulses. The pulses are then transmitted to the central processor of the computer. There are various input devices, including paper-tape readers and card readers. Some input devices, such as the visual display unit, can also be used for the output of information.

2. The process or means by which input is applied.

3. To feed information into an electrical device or machine.

See also input/output, output.

input/output (I/O) The equipment and operations used to communicate with a computer, and the information passed in or out during the communication. Input/output devices include those used only for input or for output of information and those, such as visual display units, used for both input and output. *See also* input, output.

inscribed Describing a geometric figure that is drawn inside another geometrical figure. *Compare* circumscribed.

inscribed circle *See* incircle.

instantaneous value The value of a varying quantity (e.g. velocity, acceleration, force, etc.) at a particular instant in time.

integers Symbol: z The set of whole numbers, $\{ \ldots, -2, -1, 0, 1, 2, \ldots \}$ used for counting. The integers include zero and the negative whole numbers.

integer variable *See* variable.

integral The result of integrating a function. *See* integration.

integrand A function that is to be integrated. For example, in the integral of $f(x)$. dx, $f(x)$ is the integrand. *See also* integration.

integrating factor A multiplier used to simplify and solve differential equations, usually given the symbol ξ. For example, $x \, dy - y \, dx = 2x^3 dx$ may be multiplied by $\xi(x) = 1/x^2$ to give the standard form:
$$d(y/x) = (x.dy - y.dx)/x^2 = 2x\,dx$$
which has the solution $y/x = x^2 + c$, where c is a constant. *See also* differential equation.

integration The continuous summing of change in a function, $f(x)$, over an interval of the variable x. It is the inverse process of differentiation in calculus, and its result is known as the *integral* of $f(x)$ with respect to x. An integral:
$$\int f(x)dx$$
can be regarded as the area between the curve and the x-axis, between the values x_1 and x_2. It can be considered as the sum of a number of column areas of width Δx and heights given by $f(x)$. As Δx approaches zero, the number of columns increases in-

A number line showing positive and negative integers.

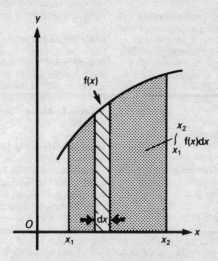

The integration of a function
$y = f(x)$ as the area between the
curve and the x-axis.

finitely and the sum of the column areas approaches the area under the curve. The integral of velocity is distance. For example, a car travelling with a velocity V in a time interval t_1 to t_2 goes a distance s given by

$$\int_{t_1}^{t_2} v\,dt$$

Integrals of this type, between definite *limits*, are known as *definite integrals*. An *indefinite integral* is one without limits. The result of an indefinite integral contains a constant – the *constant of integration*. For example,

$$\int x\,dx = x^2/2 + c$$

where c is the constant of integration. A table of integrals is given in the Appendix. *See also* differentiation.

integration by parts A method of integrating a function of a variable by expressing it in terms of two parts, both of which are differentiable functions of the same variable. A function $f(x)$ is written as the product of $u(x)$ and the derivative dv/dx. The formula for the differential of a product is:

$$d(u.v)/dx = u.dv/dx + v.du/dx$$

Integrating both sides over x and rearranging the equation gives

$$\int u.(dv/dx)dx = uv - \int v(du/dx)dx$$

which can be used to evaluate the integral of a product. For example, to integrate $x\sin x$, dv/dx is taken to be $\sin x$, so $v = -\cos x + c$ (c is a constant of integration). u is taken to be x, so $du/dx = 1$. The integral is then given by

$$\int x\sin x\,dx =$$

97

$$x(-\cos x + c) - \int(-\cos x + c) =$$
$$-x\cos x + \sin x + k$$

where k is a constant of integration. Usually a trigonometric or exponential function is chosen for dv/dx.

integration by substitution A method of integrating a function of one variable by expressing it as a simpler or more easily integrated function of another variable. For example, to integrate $\sqrt{(1 - x^2)}$ with respect to x, we can make $x = \sin u$, so that $\sqrt{(1 - x^2)} = \sqrt{(1 - \sin^2 u)} = \sqrt{(\cos^2 u)} = \cos u$, and $dx = (dx/du).du = \cos u\, du$. Therefore:

$$\int_a^b \sqrt{(1 - x^2)}dx = \int_c^d \cos^2 u.du$$

$$= [u/2 - \tfrac{1}{2}\sin u\cos u]_c^d$$

Note that for a definite integral the limits must also be changed from values of x to corresponding values of u.

intelligent terminal *See* terminal.

interaction Any mutual action between particles, systems, etc. Examples of interactions include the mutual forces of attraction between masses (gravitational interaction) and the attractive or repulsive forces between electric charges (electromagnetic interaction).

interactive terminal *See* terminal.

intercept A part of a line or plane cut off by another line or plane.

interest The amount of money paid each year at a stated rate on a borrowed capital, or the amount received each year at a stated rate on loaned or invested capital. The interest rate is usually expressed as a percentage per annum. *See* compound interest, simple interest.

interface A shared boundary. It is the area(s) or place(s) at which two devices or two systems meet and interact. There is a simple interface between an electric plug and socket. A far more complicated interface of electronic circuits provides the connection between the central processor of a computer and each of its peripheral units. Man-machine interface refers to the interaction between people and machines, including computers. For good, i.e. efficient, interaction, devices such as visual display units and easily understandable programming languages have been introduced.

interior angle An angle formed on the inside of a plane figure by two of its straight sides. For example, there are three interior angles in a triangle, which add up to 180°. *Compare* exterior angle.

internal store *See* store, central processor.

interpolation The process of estimating the value of a function from known values on either side of it. For example, if the speed of an engine, controlled by a lever, increases from 40 to 50 revolutions per second when the lever is pulled down by four centimetres, one can interpolate from this information and assume that moving it two centimetres will give 45 revolutions per second. This is the simplest method of interpolation, called *linear interpolation*. If known values of one variable y are plotted against the other variable x, an estimate of an unknown value of y can be made by drawing a straight line between the two nearest known values.
The mathematical formula for linear interpolation is:

$$y_3 = y_1 + (x_3 - x_1)(y_2 - y_1)/(x_2 - x_1)$$

y_3 is the unknown value of y (at x_3) and y_2 and y_1 (at x_2 and x_1) are the nearest known values, between which the interpolation is made. If the graph of y against x is a smooth curve, and the interval between y_1 and y_2 is small, linear interpolation can give a good approximation to the true value, but if $(y_2 - y_1)$ is large, it is less likely that y will fit sufficiently well to a straight line between y_1 and y_2. A possible source of error occurs when y is known at regular intervals, but oscillates with a period shorter than this interval.
Compare extrapolation.

interquartile range A measure of dispersion given by $(P_{75} - P_{25})$ where P_{75} is

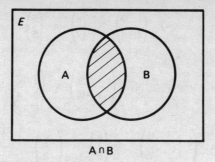

$A \cap B$

The shaded area in the Venn
diagram is the intersection of set A
and set B.

the upper quartile and P_{25} the lower quartile. The semi-interquartile range is $\frac{1}{2}(P_{75} - P_{25})$. *See also* quartile.

intersection 1. The point at which two or more lines cross each other, or a set of points that two or more geometrical figures have in common.
2. In set theory, the set formed by the elements common to two or more sets. For example, if set A is {black four-legged animals} and set B is {sheep} then the intersection of A and B, written $A \cap B$, is {black sheep}. This can be represented on a Venn diagram by the intersection of two circles, one representing A and the other B. *See* Venn diagram.

interval A set of numbers, or points in a coordinate system, defined as all the values between two end points. *See also* closed interval, open interval.

into A mapping from one set to another is said to be *into* if the range of the mapping is a proper subset of the second set; i.e. if there are members of the second set which are not the image of any element of the first set under the mapping. *Compare* onto.

intransitive Describing a relation that is not transitive; i.e. when xRy and yRz then it is not true that xRz. An example of

an intransitive relation is being the square of. If $x = y^2$ and $y = z^2$ it does not follow that $x = z^2$.

invariant Describing a property of an equation, function, or geometrical figure that is unaltered after the application of any member of some given family of transformations. For example, the area of a polygon is invariant under the group of rotations.

inverse (of a matrix) The matrix that gives as a product with another, the identity matrix I. Given a square matrix A with inverse A^{-1}, then $AA^{-1} = I$. The inverse is only defined for square matrices in which the determinant is not zero.

inverse element An element of a set that, combined with another element, gives the identity element. *See* group.

inverse function The reverse mapping of the set B into the set A when the function that maps A into B has already been defined. *See* function.

inverse hyperbolic functions The inverse functions of hyperbolic sine, hyperbolic cosine, hyperbolic tangent, etc., defined in an analogous way to the inverse trigonometric functions. For instance, the inverse hyperbolic sine of a variable x, writ-

$$\begin{aligned} x + 3y &= 5 \\ 2x + 4y &= 6 \end{aligned} \quad \Longleftrightarrow \quad \begin{pmatrix} 1 & 3 \\ 2 & 4 \end{pmatrix} \times \begin{pmatrix} x \\ y \end{pmatrix} = \begin{pmatrix} 5 \\ 6 \end{pmatrix}$$

$$\begin{pmatrix} -2 & \tfrac{3}{2} \\ 1 & -\tfrac{1}{2} \end{pmatrix} \times \begin{pmatrix} 1 & 3 \\ 2 & 4 \end{pmatrix} = \begin{pmatrix} 1 & 0 \\ 0 & 1 \end{pmatrix}$$

$$\Longrightarrow \begin{pmatrix} x \\ y \end{pmatrix} = \begin{pmatrix} -2 & \tfrac{3}{2} \\ 1 & -\tfrac{1}{2} \end{pmatrix} \times \begin{pmatrix} 5 \\ 6 \end{pmatrix} = \begin{pmatrix} -1 \\ 2 \end{pmatrix}$$

The solution of simultaneous equations by finding the inverse of a matrix. The equations are written as the equivalent matrix equation, both sides of which are then multiplied by the inverse of the coefficient matrix.

ten arc sinh$x$ or sinh^{-1}x, is the angle (or number) of which x is the hyperbolic sine. Similarly, the other inverse hyperbolic functions are:
inverse hyperbolic cosine of x (written arc cosh$x$ or cosh^{-1}x)
inverse hyperbolic tangent of x (written arc tanh$x$ or tan^{-1}x)
inverse hyperbolic cotangent of x (written arc coth$x$ or coth^{-1}x)
inverse hyperbolic secant of x (written arc sech$x$ or sech^{-1}x)
inverse hyperbolic cosecant of x (written arc cosech$x$ or cosech^{-1}x).

inverse ratio A reciprocal ratio. For example, the inverse ratio of x to y is the ratio of $1/x$ to $1/y$.

inverse square law A physical law in which an effect varies inversely as the square of the distance from the source producing the effect. An example is Newton's law of universal gravitation.

inverse trigonometric functions The inverse functions of sine, cosine, tangent, etc. For example, the *inverse sine* of a variable is called the arc sine of x; it is written arc sin$x$ or sin^{-1}x and is the angle (or number) of which the sine is x. Similar-

ly, the other inverse trigonometric functions are:
inverse cosine of x (arc cosine, written arc cos$x$ or cos^{-1}x)
inverse tangent of x (arc tangent, written arc tan$x$ or tan^{-1}x)
inverse cotangent of x (arc cotangent, written arc cot$x$ or cot^{-1}x)
inverse cosecant of x (arc cosecant, written arc cosec$x$ or cosec^{-1}x)
inverse secant of x (arc secant, written arc sec$x$ or sec^{-1}x).

inverter gate *See* logic gate.

involute The involute of a curve is a second curve that would be obtained by unwinding a taut string wrapped around the first curve. The involute is the curve traced out by the end of the string.

I/O *See* input/output.

irrational number A number that cannot be expressed as a ratio of two integers. The irrational numbers are precisely those infinite decimals that are not repeating. Irrational numbers are of two types: (i) *Algebraic irrational numbers* are roots of polynomial equations with rational numbers as coefficients. For example, $\sqrt{3} = 1.732\ 050\ 8\ldots$ is a root of the equation $x^2 = 3$. This equation does not have a rational solution since such a solution could be written $x = m/n$ with $m^2 = 3n^2$, but this is impossible since 3 divides the left-hand side an even number of times and the right-hand side an odd number of times.

(ii) *Transcendental numbers* are irrational numbers that are not algebraic, e.g. e and π.

isolated point A point that satisfies the equation of a curve but is not on the main arc of the curve. For example, the equation $y^2(x^2 - 4) = x^4$ has a solution at $x = 0$ and $y = 0$, but there is no real solution at any point near the origin, so the origin is an isolated point. *See also* double point, multiple point.

isolated system *See* closed system.

isometry A transformation in which the distances between the points remain constant.

isomorphic *See* homomorphism.

isomorphism *See* homomorphism.

isosceles Having two equal sides. *See* triangle.

issue price *See* nominal value.

iterated integral (multiple integral) A succession of integrations performed on the same function. For example, a double integral or a triple integral. *See also* double integral.

iteration A method of solving a problem by successive approximations, each using the result of the preceding approximation as a starting point to obtain a more accurate estimate. For example, the square root of 3 can be calculated by writing the equation $x^2 = 3$ in the form $2x^2 = x^2 + 3$, or $x = \frac{1}{2}(x + 3/x)$. To obtain a solution for x by iteration, we might start with a first estimate, $x_1 = 1.5$. Substituting this in the equation gives the second estimate, $x_2 = \frac{1}{2}(1.5 + 2) = 1.750\,00$. Continuing in this way, we obtain:

$$x_3 = \frac{1}{2}(1.75 + 3/1.75) = 1.732\,14$$
$$x_4 = \frac{1}{2}(1.732\,14 + 3/1.732\,14) = 1.732\,05$$

and so on, to any required accuracy. The difficulty in solving equations by iteration is in finding a formula for iteration (algorithm) that gives convergent results. In this case, for example, the algorithm $x_{n+1} = 3/x_n$ does not give convergent results. There are several standard techniques, such as Newton's method, for obtaining convergent algorithms. Iterative calculations, although often tedious for manual computation, are widely used in computers. *See also* Newton's method.

J

job A unit of work submitted to a computer. It usually includes several programs. The information necessary to run a job is input in the form of a short program written in the *job-control language* (JCL) of the computer. The JCL is interpreted by the operating system and is used to identify the job and describe its requirements to the operating system.

joule Symbol: J The SI unit of energy and work, equal to the work done where the point of application of a force of one newton moves one metre in the direction of action of the force. 1 J = 1 N m. The joule is the unit of all forms of energy.

K

kelvin Symbol: K The SI base unit of thermodynamic temperature. It is defined as the fraction $1/273.16$ of the thermodynamic temperature of the triple point of water. Zero kelvin (0 K) is absolute zero. One kelvin is the same as one degree on the Celsius scale of temperature.

Kendall's method A method of measuring the degree of association between two different ways of ranking n objects, using two variables (x and y), which give data $(x_1,y_1), \ldots ,(x_n,y_n)$. The objects are ranked using first the xs and then the ys. For each of the $2n(n-1)/2$ pairs of objects a score is assigned. If the rank of the jth object is greater (or less) than that of the kth, regardless of whether the xs or ys are used, the score is plus one. If the rank of the jth is less than that of the kth using one variable but greater using the other, the score is minus one. Kendall's coefficient of rank correlation $\tau = $ (sum of scores)$/\frac{1}{2}n(n-1)$. The closer τ is to one, the greater the degree of association between the rankings. *See also* rank, Spearman's method.

Kepler's laws Laws of planetary motion deduced in about 1610 by Johannes Kepler using astronomical observations made by Tycho Brahe:
(1) Each planet moves in an elliptical orbit with the Sun at one focus of the ellipse.
(2) The line between the planet and the Sun sweeps out equal areas in equal times.
(3) The square of the period of each planet is proportional to the cube of the semimajor axis of the ellipse.
Application of the third law to the orbit of the Moon about the Earth gave support to Newton's theory of gravitation.

keypunch *See* card.

key-to-disk *See* disk.

key-to-tape *See* magnetic tape.

kilo- Symbol: k A prefix denoting 10^3. For example, 1 kilometre (km) = 10^3 metres (m).

kilogram (kilogramme) Symbol: kg The SI base unit of mass, equal to the mass of the international prototype of the kilogram, which is a piece of platinum–iridium kept at Sèvres in France.

kilogramme An alternative spelling of *kilogram*.

kilowatt-hour Symbol: kwh A unit of energy, usually electrical, equal to the energy transferred by one kilowatt of power in one hour. It is the same as the Board of Trade unit and has a value of 3.6×10^6 joules.

kinematics The study of the motion of objects without consideration of its cause. *See also* mechanics.

kinetic energy Symbol: T The work that an object can do because of its motion. For an object of mass m moving with velocity v, the kinetic energy is $mv^2/2$. This gives the work the object would do in coming to rest. The rotational kinetic energy of an object of moment of inertia I and angular velocity ω is given by $I\omega^2/2$. *See also* energy.

kinetic friction *See* friction.

kite A plane figure with four sides, and two pairs of adjacent sides equal. Two of the angles in a kite are opposite and equal. Its diagonals cross perpendicularly, one of them (the shorter one) being bisected by the other. The area of a kite is equal to the product of its diagonal lengths. In the special case in which the two diagonals have equal lengths, the kite is a rhombus.

Klein bottle A curved surface with the unique topological property that it has only one surface, no edges, and no inside

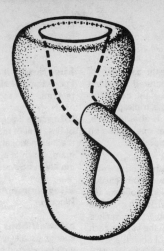

The Klein bottle, a single closed
surface with no inside.

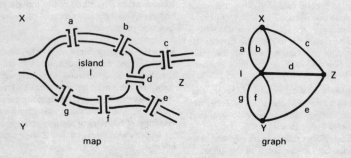

map   graph

Königsberg bridge problem

and outside. It can be thought of as
formed by taking a length of flexible
stretchy tubing, cutting a hole in the side
through which one end can fit exactly,
passing an end of the tube through this
hole, and then joining it to the other end
from the inside. Starting at any point on
the surface a continuous line can be drawn
along it to any other point without crossing
an edge. *See also* topology.

knot A curve formed by looping and in-
terlacing a string and then joining the
ends. The mathematical theory of knots is
a branch of topology. *See also* topology.

knot A unit of velocity equal to one nauti-
cal mile per hour. It is equal to 0.414
m s⁻¹.

Königsberg bridge problem A classical problem in topology. The river in the Prussian city of Königsberg divided into two branches and was crossed by seven bridges in a certain arrangement. The problem was to show that it is impossible to walk in a continuous path across all the bridges and cross each one only once. The problem was solved by Euler in the eight-eenth century, by replacing the arrange-ment by an equivalent one of lines and vertices. He showed that a network like this (called a *graph*) can be traversed in a single path if and only if there are fewer than three vertices at which an odd number of line segments meet. In this case there are four.

L

language Short for *programming language. See* program.

Laplace equation *See* partial differential equation.

laser printer A type of computer printer in which the image is formed by scanning a charged plate with a laser. Powdered ink adheres to the charged areas and is transferred to paper as in a photocopier. Laser printers can produce text almost as good as typesetting machines.

lateral Denoting the side of a solid geometrical figure, as opposed to the base. For instance, a lateral edge of a pyramid is one of the edges from the vertex (apex). A lateral face of a pyramid or prism is a face that is not a base. The lateral surface (or area) of a cylinder or cone is the curved surface (or area), excluding the plane base.

latin square An $n \times n$ square array of n different symbols with the property that each symbol appears once and only once in each row and each column. Such an arrangement is possible for every n. For example, for $n = 4$ and the letters A, B, C, D:

A B C D
C D A B
D C B A
B A D C

Such arrays are used in statistics to analyse experiments with three factors influencing the outcome; for example the experimenter, the method, and the material under test. The significance of each factor in the experiment may be tested by using a latin square. For example, denoting the four methods by A, B, C, D we may use the latin square

material
experimenter
A B C D
C D A B
D C B A
B A D C

The latin-square array is used since, in studying the effects of one factor, the influences of the other factors occur to the same extent.

latitude The distance of a point on the Earth's surface from the equator, measured as the angle in degrees between a plane through the equator (the equatorial plane) and the line from the point to the centre of the Earth. A point on the equator has a latitude of $0°$ and the North Pole has a latitude of $90°$. *See also* longitude.

lattice A partially ordered set such that each pair of elements a and b has both:
(1) A greatest lower bound c; i.e. an element c such that $c \leqslant a$ and $c \leqslant b$ and if $c' \leqslant a$ and $c' \leqslant b$ then $c' \leqslant c$.
(2) A least upper bound d; i.e. an element d such that $d \geqslant a$ and $d \geqslant b$ and if $d' \geqslant a$ and $d' \geqslant b$ then $d' \geqslant d$.
The elements c and d are called the *meet* and *join* respectively of a and b and are denoted by $c = a \cap b$ and $d = a \cup b$. An example of a lattice is the set of all subsets of a given set, where $A \leqslant B$ means that each element of A is also an element of B. In this example, $A \cap B$ is the intersection of the sets A and B and $A \cup B$ is their union.

latus rectum *See* ellipse, hyperbola, parabola.

law of flotation *See* flotation, law of.

law of large numbers A theorem in probability stating that if an event E has probability p, and if $N(E)$ represents the number of occurrences of the event in n trials, then $N(E)/n$ is very close to p if n is a large number and $N(E)/n$ converges to p as n tends to infinity.

106

law of moments *See* moment.

law of the mean (mean value theorem) The rule in differential calculus that, if f(x) is continuous in the interval $a \leqslant x \leqslant b$, and the derivative f'(x) exists everywhere in this interval, then there is at least one value of x (x_0) between a and b for which:

$$[f(b) - f(a)]/(b - a) = f'(x_0)$$

Geometrically this means that if a straight line is drawn between two points, (a,f(a)) and (b,f(b)), on a continuous curve, then there is at least one point between these where the tangent to the curve is parallel to the line. This law is derived from Rolle's theorem. *See also* Rolle's theorem.

laws of conservation *See* conservation law.

laws of friction *See* friction.

laws of thought Three laws of logic that are traditionally considered – as are other logic rules – to exemplify something fundamental about the way we think; that is, it is not arbitrary that we say certain forms of reasoning are correct. On the contrary, it would be impossible to think otherwise.

1. *The law of contradiction* (law of non-contradiction). It is not the case that something can both be true and not be true: symbolically

$$\sim(p \wedge \sim p)$$

2. *The law of excluded middle.* Something must either be true or not be true: symbolically

$$p \vee \sim p$$

3. *The law of identity.* If something is true, then it is true: symbolically

$$p \to p$$

LCD Lowest common denominator. *See* common denominator.

LCM Lowest common multiple. *See* common multiple.

leading diagonal *See* square matrix.

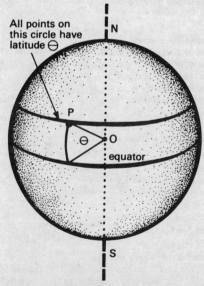

All points on this circle have latitude Θ

The latitude Θ of a point P on the Earth's surface.

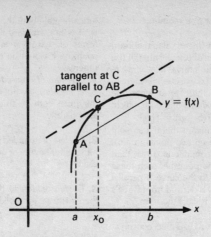

The law of the mean for a function
f(x) that is continuous between
$x = a$ and $x = b$.

least common denominator *See*
common denominator.

least common multiple *See* common multiple.

least squares method A method of
fitting a regression line to a set of data. If
the data are points $(x_1, y_1), \ldots (x_m, y_n)$, the
corresponding points $(x_1, y_1'), \ldots (x_m, y_n')$
are found using the linear equation $y = ax + b$. The least squares line minimizes
$$[(y_1 - y_1')^2 + (y_2 - y_2')^2 + \ldots + (y_n - y_n')^2]$$
It is found by solving the normal equations
$$\Sigma y = an + b\Sigma x$$
$$\Sigma xy = a\Sigma x + b\Sigma x^2$$
for a and b. The technique is extended for
regression quadratics, cubics, etc. *See also* regression line.

least upper bound *See* bound.

Legendre polynomial A series of
functions that arises as solutions to
Laplace's equation in spherical polar coordinates. They form an infinite series. *See
also* partial differential equation.

Leibniz theorem A formula for finding
the nth derivative of a product of two functions. The nth derivative with respect to x
of a function $f(x) = u(x)v(x)$, written $D^n(uv)$
$= d^n(uv)/dx^n$, is equal to the series:
$$uD^n v + {}_nC_1 Du D^{n-1} v + {}_nC_2 D^2 u D^{n-2} v +$$
$$\ldots + {}_nC_{n-1} D^{n-1} u Dv + v D^n u$$
where ${}_nC_r = n!/[(n-r)!r!]$.
The formula holds for all positive integer
values of n.
For $n = 1$, $D(uv) = uDv + vDu$
For $n = 2$, $D^2(uv) = uD^2 v + 2DuDv + vD^2 u$
For $n = 3$, $D^3(uv) = uD^3 v + 3DuD^2 v + 3D^2 uDv + D^3 u$
Note the similarity between the differential
coefficients and the binomial expansion
coefficients.

lemma A theorem proved for use in the
proof of another theorem. *See* theorem.

length The distance along a line, plane
figure, or solid. In a rectangle, the greater
of the two dimensions is usually called the
length and the smaller the breadth.

lever A class of machine; a rigid object able to turn around some point (pivot or fulcrum). The force ratio and the distance ratio depend on the relative positions of the fulcrum, the point where the user exerts the effort, and the point where the lever applies force to the load. There are three types (orders) of lever.
First order, in which the fulcrum is between the load and the effort. An example is a crowbar.
Second order, in which the load is between the effort and the fulcrum. An example is a wheelbarrow.
Third order, in which the effort is between the load and the fulcrum. An example is a pair of sugar tongs.
Levers can have high efficiency; the main energy losses are by friction at the pivot, and by bending of the lever itself. *See* machine.

library programs Collections of computer programs that have been purchased, contributed by users, or supplied by the computer manufacturers for the use of the computing community. Libraries that have been purchased or supplied by manufacturers usually have a theme, such as engineering or mathematics.

light-year Symbol: ly A unit of distance used in astronomy, defined as the distance that light travels through space in one year. It is approximately equal to $9.460\,5 \times 10^{15}$ metres.

limit In general, the value approached by a function as the independent variable approaches some value. The idea of a limit is the basis of the branch of mathematics known as *analysis*. There are several examples of the use of limits.
(1) The limit of a function is the value it approaches as the independent variable tends to some value or to infinity. For instance, the function $x/(x+3)$ for positive values of x is less than 1. As x increases it approaches 1 – the value approached as x becomes infinite. This is written

$$\text{Lim } x/(x+3) = 1$$
$$x \to \infty$$

stated as 'the limit of $x/(x+3)$ as x approaches (or tends to) infinity is 1'. 1 is the *limiting value* of the function.
(2) The limit of a convergent sequence is the limit of the nth term as n approaches infinity. *See* convergent sequence.
(3) The limit of a convergent series is the limit of the sum of n terms as n approaches infinity. *See* convergent series.
(4) A derivative of a function $f(x)$ is the limit of $[f(x+\delta x) - f(x)]/\delta x$ as δx approaches zero. *See* derivative.
(5) A definite integral is the limit of a finite sum of terms $y\delta x$ as δx approaches zero. *See* integral.

limiting friction *See* friction.

line A join between two points in space or on a surface. A line has length but no breadth, that is, it has only one dimension. A straight line is the shortest distance between two points on a flat surface.

linear dependence *See* dependent.

linear equation An equation in which the highest power of an unknown variable is one. The general form of a linear equation is

$$mx + c = 0$$

where m and c are constants. On a Cartesian coordinate graph

$$y = mx + c$$

is a straight line that has a gradient m and crosses the y-axis at $y = c$. The equation

$$x + 4y^2 = 4$$

is linear in x but not in y. *See also* equation.

linear extrapolation *See* extrapolation.

linear independence *See* dependent.

linear interpolation *See* interpolation.

linearly ordered A *linearly ordered set* is a partially ordered set that satisfies the *trichotomy principle*: for any two elements x and y exactly one of $x < y, x > y, x = y$ is true. For example, the set of positive integers with their natural order is a linearly ordered set.

linear momentum *See* momentum.

linear momentum, conservation of *See* constant linear momentum, law of.

linear programming The process of finding maximum or minimum values of a linear function under limiting conditions or constraints. For example, the function $x - 3y$ might be minimized subject to the constraints that $x + y \leqslant 10$, $x \leqslant y$, $x \geqslant 0$ and $y \geqslant 0$. The constraints can be shown as the area on a Cartesian coordinate graph bounded by the lines $x + y = 10$; $x = y$, $x = 0$, and $y = 0$. The minimum value for $x - 3y$ is chosen from points within this area. A series of parallel lines $x - 3y = k$ are drawn for different values of k. The line $k = -9$ just reaches the constraint area at the point $(10,0)$. Lower values are outside it, and so $x = 10$, $y = 0$ gives the minimum value of $x - 3y$ within the constraints. Linear programming is used to find the best possible combination of two or more variable quantities that determine the value of another quantity. In most applications, for example, finding the best combination of quantities of each product from a factory to give the maximum profit, there are many variables and constraints. Linear functions with large numbers of variables and constraints are maximized or minimized by computer techniques that are similar in principle to this graphical technique for two variables.

linear transformation 1. In one dimension, the general linear transformation is given by:
$$x' = (ax + b)/(dx + c)$$
where a, b, c, and d are constants. In two dimensions, the general linear transformation is given by:
$$x' = (a_1x + b_1y + c_1)/(d_1x + e_1y + f_2)$$
and
$$y' = (a_2x + b_2y + c_2)/(d_2x + e_2y + f_2)$$
General linear transformations in more than two dimensions are defined similarly.

2. In an n-dimensional vector space, a linear transformation has the form $y = Ax$, where x and y are column vectors and A is a matrix. A linear transformation takes $ax + by$ into $ax' + by'$ for all a and b if it takes x and y into x' and y'.

line integral (contour integral, curvilinear integral) The integration of a function along a particular path, C, which may be a segment of straight line, a portion of space curve, or connected segments of several curves. The function is integrated with respect to the position vector $r = ix + jy + kz$, which denotes the position of each point $P(x,y,z)$ on a curve C.

For example, the direction and magnitude of a force vector F acting on a particle may depend on the particle's position in a gravitational field or a magnetic field. The work done by the force in moving the particle over a distance dr is $F.dr$. The total work done in moving the particle along a particular path from point P_1 to point P_2 is the line integral shown in the diagram.

line printer An output device of a computer system that prints characters (letters, numbers, punctuation marks, etc.) on paper a complete line at a time and can therefore operate very rapidly; 1000 lines per minute is a typical speed. The *barrel* (or *drum*) *printer* is widely used. It has a set of printing characters embossed around the circumference of a barrel at each character position across the page. The paper is continuous, with a line of perforations between each sheet and sprocket holes along each side to control its movement.

Lissajous figures Patterns obtained by combining two simple harmonic motions in different directions. For example, an object moving in a plane so that two components of the motion at right angles are simple harmonic motions, traces out a Lissajous figure. If the components have the same frequency and amplitude and are in phase the motion is a straight line. If they are out of phase by 90°, it is a circle. Other phase differences produce ellipses. If the frequencies of the components dif-

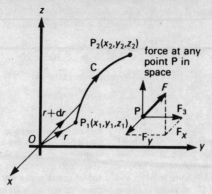

The line integral of a force vector F along a path C from a point P_1 to a point P_2.

$$\int_{P_1}^{P_2} F.dr = \int_{x_1}^{x_2} F_x d + \int_{y_1}^{y_2} F_y\, dy + \int_{z_1}^{z_2} F_z dz$$

fer, more complex patterns are formed. Lissajous figures can be demonstrated with an oscilloscope by deflecting the spot with one oscillating signal along one axis and with another signal along the other axis.

litre Symbol: l A unit of volume defined as 10^{-3} metre^3. The name is not recommended for precise measurements. Formerly, the litre was defined as the volume of one kilogram of pure water at 4°C and standard pressure. On this definition, 1 l = 1000.028 cm^3.

load The force generated by a machine. *See* machine.

local maximum (relative maximum) A value of a function f(x) that is greater than for the adjacent values of x, but is not the greatest of all values of x. *See* maximum point.

local meridian *See* longitude.

local minimum (relative minimum) A value of a function f(x) that is less than for

the adjacent values of x but is not the lowest of all values of x. *See* minimum point.

location *See* store.

locus of points A set of points, often defined by an equation relating coordinates. For example, in rectangular Cartesian coordinates, the locus of points on a line through the origin at 45° to the x-axis and the y-axis is defined by the equation x = y; the line is said to be the locus of the equation. A circle is the locus of all points that lie a fixed distance from a given point.

logarithm A number expressed as the exponent of another number. Any number x can be written in the form $x = a^y$. y is then the logarithm to the base a of x. For example, the logarithm to the base ten of 100 ($\log_{10}100$) is two, since $100 = 10^2$. Logarithms that have a base of ten are known as *common logarithms* (or *Briggsian logarithms*). They are used to carry out multiplication and division calculations because numbers can be multiplied by adding their logarithms. In general p × q can be written as $a^c \times a^d = a^{(c+d)}$, $p = a^c$

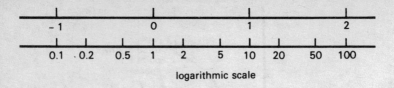

logarithmic scale

and $q = a^d$. Both logarithms and anti-logarithms (the inverse function) are available in the form of printed tables. 4.91×5.12 would be calculated as follows: $\log_{10}4.91$ is 0.6911 (from tables) and $\log_{10}5.12$ is 0.7093 (from tables). Therefore, 4.91×5.12 is given by antilog (0.6911 + 0.7093) being 25.14 (from antilog tables). Similarly, division can be carried out by subtraction of logarithms and the nth root of a number (x) is the antilogarithm of $(\log x)/n$.

For numbers between 0 and 1 the common logarithm is negative. For example, $\log_{10}0.01 = -2$. The common logarithm of any positive real number can be written in the form $n + \log_{10}x$, where x is between 1 and 10 and n is an integer. For example,
$\log_{10}15 = \log_{10}(10 \times 1.5) = \log_{10}10 + \log_{10}1.5 = 1 + 0.1761$
$\log_{10}150 = \log_{10}(100 \times 1.5) = 2.1761$
$\log_{10}0.15 = \log_{10}(0.1 \times 1.5) = -1 + 0.1761$. This is written in the notation $\bar{1}.1761$.

The integer part of the logarithm is called the *characteristic* and the decimal fraction is the *mantissa*. *Natural logarithms* (*Napierian logarithms*) use the base $e = 2.71828\ldots$, $\log_e x$ is often written as $\ln x$.

logarithmic function The function $\log_a x$, where a is a constant. It is defined for positive values of x.

logarithmic scale 1. A line in which the distance, x, from a reference point is proportional to the logarithm of a number. For example, one unit of length along the line might represent 10, two units 100, three units 1000, and so on. In this case the distance x along the logarithmic scale is given by the equation $x = \log_{10}a$. Logarithmic scales form the basis of the slide rule since two numbers may be multiplied by adding lengths on a logarithmic scale ($\log(a \times b) = \log a + \log b$.

The graph of the curve $y = x^n$, when plotted on graph paper with logarithmic coordinate scales on both axes (log-log graph paper), is a straight line since $\log y = n\log x$. This method can be used to establish the equation of a non-linear curve. Known values of x and y are plotted on log-log graph paper and the gradient n of the resulting line is measured, enabling the equation to be found. *See also* log-linear graph.

2. Any scale of measurement that varies logarithmically with the quantity measured. For instance, pH in chemistry is a measure of acidity or alkalinity – i.e. of hydrogen-ion concentration. It is defined as $\log_{10}(1/[H^+])$. An increase in pH from 5 to 6 represents a decrease in $[H^+]$ from 10^{-5} to 10^{-6}, i.e. a factor of 10. An example of a logarithmic scale in physics is the decibel scale used for noise level.

logarithmic series The infinite power series that is the expansion of the function $\log_e(1+x)$, namely:
$$x - x^2/2 + x^3/3 - x^4/4 + \ldots$$
This series is convergent for all values of x greater than -1 and less than or equal to 1.

logic The study of the methods and principles used in distinguishing correct or valid arguments and reasoning from incorrect or invalid arguments and reasoning. The main concern in logic is not whether a conclusion is in fact accurate, but whether the process by which it is derived from a set of initial assumptions (*premisses*) is correct. Thus, for example, the following form of argument is valid:

input 1	input 2	output
high	high	low
high	low	high
low	high	high
low	low	low

Table for a logic circuit

all *A* is *B*
all *B* is *C*
therefore all *A* is *C*,
and thus the conclusion
all fish have wings
can be derived, correctly, from the premisses
all fish are mammals
and
all mammals have wings
even though the premisses and the conclusion are untrue. Similarly, true premisses and true conclusions are no guarantee of a valid argument. The conclusion
all cats are mammals
does *not* follow, logically, from the true premisses:
all cats are warm-blooded
and
all mammals are warm blooded
because it is an example of the *invalid* argument form
all *A* is *B*
all *C* is *B*
therefore, all *A* is *C*.
The incorrectness of the argument shows up clearly when, after making reasonable substitutions for *A*, *B*, and *C*, we get *true* premisses but a *false* conclusion:
all dogs are mammals
all cats are mammals
therefore all dogs are cats.
Such an argument is called a *fallacy*.
Logic puts forward and examines rules that will ensure that – given true premisses – a true conclusion can automatically be reached. It is not concerned with examining or assessing the truth of the premisses; it is concerned with the form and structure of arguments, not their content. *See* deduction, induction, symbolic logic, truth-value, validity.

logic circuit An electronic switching circuit that performs a logical operation, such as 'and' and 'implies' on its input signals. There are two possible levels for the input and output signals, high and low, sometimes indicated by the binary digits 1 and 0, which can be combined like the values 'true' and 'false' in a truth table. For example, a circuit with two inputs and one output might have a high output only when the inputs are different. The output therefore is the logical function 'either . . . or . . . ' of the two inputs (the exclusive disjunction). *See* truth table.

logic gate An electronic circuit that carries out logical operations. Examples of such operations are 'and', 'either – or', 'not', 'neither – nor', etc. Logic gates operate on high or low input and output voltages. Binary logic circuits, those that switch between two voltage levels (high and low), are widely used in digital computers. The *inverter gate* or *NOT gate* simply changes a high input to a low output and vice versa. In its simplest form, the *AND gate* has two inputs and one output. The output is high if and only if both inputs are high. The *NAND gate* (not and) is similar, but has the opposite effect; that is, a low output if and only if both inputs are high. The *OR gate* has a high output if one or more of the inputs are high. The *exclusive OR gate* has a high input only if one of the inputs, but not more than one, is high. The *NOR gate* has a high output only if all the inputs are low. Logic gates are constructed using transistors, but in a circuit diagram they are often shown by symbols that denote only their logical functions. These functions are, in effect, those relationships that can hold between propositions in symbolic logic, with combinations

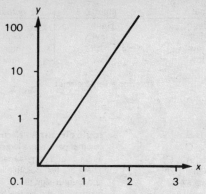

In this log-linear graph, the function $y = 4.9\,e^{1\cdot5x}$ is shown as a straight line with a gradient of 1.5.

that can be represented in a truth table. *See also* conjunction, disjunction, negation, truth table.

log-linear graph (semilogarithmic graph) A graph on which one axis has a logarithmic scale and the other has a linear scale. On a log-linear graph, an exponential function (one of the form $y = ke^{ax}$ where k and a are constants) is a straight line. Values of y are plotted on the linear scale and values of x on the logarithmic scale. *See also* logarithmic scale.

log-log graph A graph on which both axes have logarithmic scales. *See* logarithmic scale.

longitude The east-west position of a point on the Earth's surface measured as the angle in degrees from a standard meridian (taken as the Greenwich meridian). A *meridian* is a great circle that passes through the North and South poles. The *local meridian* of a point is a great circle passing through that point and the two poles. *See also* latitude.

longitudinal wave A wave motion in which the vibrations in the medium are in the same direction as the direction of energy transfer. Sound waves, transmitted by alternate compression and rarefaction of the medium, are an example of longitudinal waves. *Compare* transverse wave.

loop A sequence of instructions in a computer program that is performed either a specified number of times or is performed repeatedly until some condition is satisfied. *See also* branch.

Lorentz–Fitzgerald contraction A reduction in the length of a body moving with a speed v relative to an observer, as compared with the length of an identical object at rest relative to the observer. The object is supposed to contract by a factor $\sqrt{(1 - v^2/c^2)}$, c being the speed of light in free space. The contraction was postulated to account for the negative result of the Michelson–Morley experiment using the ideas of classical physics. The idea behind it was that the electromagnetic forces holding atoms together were modified by motion through the ether. The idea was made superfluous (along with the concept of the ether) by the theory of relativity, which supplied an alternative explanation of the Michelson–Morley experiment.

lower bound *See* bound.

lowest common denominator *See* common denominator.

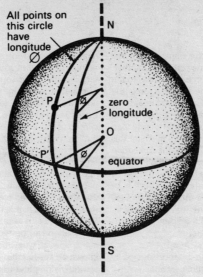

All points on this circle have longitude Ø

N

zero longitude

P

O

P'

equator

S

The longitude Ø of a point P on the Earth's surface.

lowest common multiple *See* common multiple.

low-level language *See* program.

lumen Symbol: lm The SI unit of luminous flux, equal to the luminous flux emitted by a point source of one candela in a solid angle of one steradian. 1 lm = 1 cd sr.

lune A portion of the area of a sphere bounded by two great semicircles that have common end points.

lux Symbol: lx The SI unit of illumination, equal to the illumination produced by a luminous flux of one lumen falling on a surface of one square metre. $1 \, \text{lx} = 1 \, \text{lm} \, \text{m}^{-2}$.

machine A device for transmitting force or energy between one place and another. The user applies a force (the effort) to the machine; the machine applies a force (the load) to something. These two forces need not be the same; in fact the purpose of the machine is often to overcome a large load with a small effort. For any machine this relationship is measured by the *force ratio* (or mechanical advantage) -- the force applied by the machine (load, F_2) divided by the force applied by the user (effort, F_1).

The work done by the machine cannot exceed the work done to the machine. Therefore, for a 100% efficient machine:
if $F_2 > F_1$ then $s_2 < s_1$
and if $F_2 < F_1$ then $s_2 > s_1$.
Here s_2 and s_1 are the distances moved by F_2 and F_1 in a given time.

The relationship between s_1 and s_2 in a given case is measured by the *distance ratio* (or velocity ratio) -- the distance moved by the effort (s_1) divided by the distance moved by the load (s_2).

Neither distance ratio nor force ratio have a unit; neither has a standard symbol. *See also* hydraulic press, inclined plane, lever, pulley, screw, wheel and axle.

machine code *See* program.

machine language *See* program.

Maclaurin series *See* Taylor series.

magnetic disk *See* disk.

magnetic drum *See* drum.

magnetic tape A long strip of flexible plastic with a magnetic coating on which information can be stored. Its use in the recording and reproduction of sound is well-known. It is also widely used in computing to store information. The data is stored on the tape in the form of small closely packed magnetic spots arranged in rows across the tape. The spots are magnetized in one of two directions so that the data is in binary form. The magnetization pattern of a row of spots represents a letter, digit (0–9), or some other character. There are usually nine or seven spot positions across a tape, forming columns or *tracks* along its length. A number of adjacent rows is used to store an item of information. With 800, 1600, or 6250 rows per inch of tape, one magnetic tape can store an immense amount of information. This information can be altered or deleted as necessary by magnetic means. A tape can therefore be reused many times. The tape must be of very high quality and is typically ½ inch (1.27 cm) wide.

Information can be recorded on tape using a special typewriter; this method is known as *key-to-tape*. The information is fed into a computer using a *magnetic tape unit* (MTU). In the simplest version two tape reels are driven at very high speed so that a magnetic tape is wound from one reel to the other and back again. As it does so each track on the tape passes close to a small electromagnet, called a *read-write head*. These extract (read) information required by the central processor of the computer, according to program instructions, and record (write) information that is sent from the central processor. A particular piece of information can only be read or written when the tape has been wound into position under the heads. The MTU is therefore a serial-access as opposed to a random-access device. It is widely used as a backing store. *Compare* card, disk, drum.

magnitude 1. The absolute value of a number (without regard to sign).
2. The non-directional part of a vector, corresponding to the length of the line representing it. *See* vector.

mainframe *See* computer.

main store *See* store, central processor.

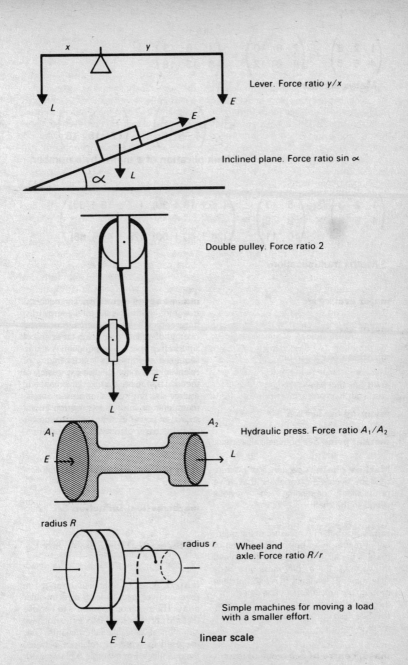

Lever. Force ratio y/x

Inclined plane. Force ratio $\sin \alpha$

Double pulley. Force ratio 2

Hydraulic press. Force ratio A_1/A_2

Wheel and axle. Force ratio R/r

Simple machines for moving a load with a smaller effort.

linear scale

$$\begin{pmatrix} 1 & 2 & 3 \\ 4 & 5 & 6 \end{pmatrix} + \begin{pmatrix} 2 & 6 & 10 \\ 4 & 8 & 12 \end{pmatrix} = \begin{pmatrix} 3 & 8 & 13 \\ 8 & 13 & 18 \end{pmatrix}$$

Matrix addition.

$$3 \times \begin{pmatrix} 1 & 2 & 3 \\ 4 & 5 & 6 \end{pmatrix} = \begin{pmatrix} 3 & 6 & 9 \\ 12 & 15 & 18 \end{pmatrix}$$

Multiplication of a matrix by a number.

$$\begin{pmatrix} 1 & 2 & 3 \\ 4 & 5 & 6 \end{pmatrix} \times \begin{pmatrix} 6 & 7 \\ 8 & 9 \\ 10 & 11 \end{pmatrix} = \begin{pmatrix} (6+16+30) & (7+18+33) \\ (24+40+60) & (28+45+66) \end{pmatrix}$$

Matrix multiplication.

major arc *See* arc.

major axis *See* ellipse.

mantissa *See* logarithm.

map *See* function.

mapping *See* function.

market price *See* nominal value, yield.

Markov chain A sequence of discrete random events or variables that have probabilities depending on previous events in the chain.

mass Symbol: m A measure of the quantity of matter in an object. The SI unit of mass is the kilogram. Mass is determined in two ways: the *inertial mass* of a body determines its tendency to resist change in motion; the *gravitational mass* determines its gravitational attraction for other masses. *See also* gravitational mass, inertial mass, weight.

mass, centre of *See* centre of mass.

mass-energy equation The equation $E = mc^2$, where E is the total energy (rest mass energy + kinetic energy + potential energy) of a mass m, c being the speed of light in free space. The equation is a consequence of Einstein's special theory of relativity. It is a quantitative expression of the idea that mass is a form of energy and energy also has mass. Conversion of rest-mass energy into kinetic energy is the source of power in radioactive substances and the basis of nuclear-power generation.

material implication *See* implication.

mathematical induction *See* induction.

mathematical logic *See* symbolic logic.

matrix A set of quantities arranged in rows and columns to form a rectangular array. The common notation is to enclose these in parentheses. Matrices do not have a numerical value, like determinants. They are used to represent relations between the quantities. For example, a plane vector

can be represented by a single column matrix with two numbers, a 2×1 matrix, in which the upper number represents its component parallel to the x-axis and the lower number represents the component parallel to the y-axis. Matrices can also be used to represent, and solve, simultaneous equations. In general, an $m \times n$ matrix − one with m rows and n columns − is written with the first row:
$$A = a_{11}a_{12} \ldots a_{1n}$$
The second row is:
$$a_{21}a_{22} \ldots a_{2n}$$
and so on, the mth row being:
$$a_{m1}a_{m2} \ldots a_{mn}$$
The individual quantities a_{11}, a_{21}, etc., are called *elements* of the matrix. The number of rows and columns, $m \times n$, is the *order* or *dimensions* of the matrix. Two matrices are equal only if they are of the same order and if all their corresponding elements are equal. Matrices, like numbers, can be added, subtracted, multiplied, and treated algebraically according to certain rules. However, the commutative, associative, and distributive laws of ordinary arithmetic do not apply. *Matrix addition* consists of adding corresponding elements together to obtain another matrix of the same order. Only matrices of the same order can be added. Similarly, the result of subtracting two matrices is the matrix formed by the differences between corresponding elements.

Matrix multiplication also has certain rules. In multiplication of an $m \times n$ matrix by a number or constant k, the result is another $m \times n$ matrix. If the element in the ith row and jth column is a_{ij} then the corresponding element in the product is ka_{ij}. This operation is distributive over matrix addition and subtraction, that is, for two matrices A and B,
$$k(A + B) = kA + kB$$
Also, $kA = Ak$, as for multiplication of numbers. In the multiplication of two matrices, the matrices A and B can only be multiplied together to form the product AB if the number of columns in A is the same as the number of rows in B. In this case they are called *conformable matrices*. If A is an $m \times p$ matrix with elements a_{ij} and B is a $p \times n$ matrix with elements b_{ij},

then their product $AB = C$ is an $m \times n$ matrix with elements c_{ij}, such that c_{ij} is the sum of the products
$$a_{i1}b_{1j} + a_{i2}b_{2j} + a_{i3}b_{3j} + \ldots + a_{ip}b_{pj}$$
Matrix multiplication is not commutative, that is, $AB \neq BA$.
See also determinant, square matrix.

maximum likelihood A method of estimating the most likely value of a parameter. If a series of observations $x_1, x_2, \ldots, x_n$ are made, the likelihood function, $L(x)$, is the joint probability of observing these values. The likelihood function is maximized when $[d\log L(x)]/dp = 0$. In many cases, an intuitive estimate, such as the mean, is also the maximum likelihood estimate.

maximum point A point on the graph of a function at which it has the highest value within an interval. If the function is a smooth continuous curve, the maximum is a turning point, that is, the slope of the tangent to the curve changes continuously from positive to negative by passing through zero. If there is a higher value of the function outside the immediate neighbourhood of the maximum, it is a *local maximum* (or *relative maximum*). If it is higher than all other values of the function it is an *absolute maximum*. *See also* stationary point, turning point.

mean A representative or expected value for a set of numbers. The *arithmetic mean* or *average* (called *mean*) of $x_1, x_2, \ldots, x_n$ is given by:
$$(x_1 + x_2 + x_3 + \ldots + x_n)/n$$
If $x_1, x_2, \ldots, x_k$ occur with frequencies $f_1, f_2, \ldots, f_k$ then the arithmetic mean is
$$(f_1 x_1 + f_2 x_2 + \ldots + f_k x_k)/$$
$$(f_1 + f_2 + \ldots + f_k)$$
When data is classified, as in a frequency table, x_1 is replaced by the class mark. The *weighted mean* $W =$
$$(w_1 x_1 + w_2 x_2 + \ldots + w_n x_n)/$$
$$(w_1 + w_2 + \ldots + w_n)$$
where weight w_i is associated with x_i. The *harmonic mean* $H =$
$$n/[(1/x_1) + (1/x_2) + \ldots + (1/x_n)]$$
The *geometric mean* $G =$
$$(x_1 . x_2 \ldots x_n)^{1/n}$$

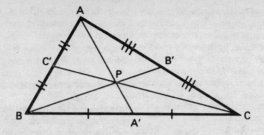

The three medians AA', BB', and CC' intersect at a point P, the centroid of the triangle.

The mean of a random variable is its expected value.

mean centre *See* centroid.

mean deviation A measure of the dispersion of a set of numbers. It is equal to the average, for a set of numbers, of the differences between each number and the set's mean value. If x is a random variable with mean value μ, then the mean deviation is the average, or expected value, of lx − μl, written

$$\Sigma lx_i - \mu l / n$$

mean value theorem *See* law of the mean.

mechanical advantage *See* force ratio.

mechanics The study of forces and their effect on objects. If the forces on an object or in a system cause no change of momentum the object or system is in equilibrium. The study of such cases is *statics*. If the forces acting do change momentum the study is of *dynamics*. The ideas of dynamics relate the forces to the momentum changes produced. *Kinematics* is the study of motion without consideration of its cause.

median 1. The middle number of a set of numbers arranged in order. When there is an even number of numbers the median is the average of the middle two. For example, the median of 1,3,5,11,11 is 5, and of 1,3,5,11,11,14 is (5 + 11)/2 = 8. The median of a large population is the 50th percentile (P_{50}). *Compare* mean. *See also* percentile, quartile.
2. In geometry, a straight line joining the vertex of a triangle to the mid-point of the opposite side. The medians of a triangle intersect at a single point, which is the centroid of the triangle.

mega- Symbol: M A prefix denoting 10^6. For example, 1 megahertz (MHz) = 10^6 hertz (Hz).

member *See* element.

memory *See* store.

mensuration The study of measurements, especially of the dimensions of geometric figures in order to calculate their areas and volumes.

Mercator's projection A method of mapping points from the surface of a sphere onto a plane surface. Mercator's projection is used to make maps of the World. The lines of longitude on the sphere become straight vertical lines on the plane. The lines of latitude on the sphere become straight horizontal lines. Areas further from the equator are more stretched out in the horizontal direction. For a particular point on the surface of the sphere at angle θ latitude and angle ϕ longitude, the corresponding Cartesian coordinates on the map are:

$$x = k\theta$$
$$y = k \log \tan(\phi/2)$$

Mercator's projection is an example of a conformal mapping, in which the angles between lines are preserved (except at the poles).
See also projection.

meridian *See* longitude.

metre Symbol: m The SI base unit of length, defined as the length equal to 1 650 763.73 wavelengths in vacuum corresponding to the transition between the levels $2p_{10}$ and $5d_2$ of the krypton-86 atom.

metric space Any set of points, such as a plane or a volume in geometrical space, in which a pair of points a and b with a distance $d(a,b)$ between them, satisfy the conditions that $d(a,b) \geqslant 0$ and $d(a,b) = 0$ if and only if a and b are the same point. Another property of a metric space is that $d(a,b) + d(b,c) \geqslant d(a,c)$. The set of all the functions of x that are continuous in the interval $x = a$ to $x = b$ is also a metric space. If $f(x)$ and $g(x)$ are in the space,

$$\int_a^b [f(x) - g(x)]dx$$

is defined for all values of x between a and b, and the integral is zero if and only if $f(x) = g(x)$ for all values of x between a and b.

metric system A system of units based on the metre and the kilogram and using multiples and submultiples of 10. SI units, c.g.s. units, and m.k.s. units are all scientific metric systems of units.

metric ton *See* tonne.

metrology The study of units of measurements and the methods of making precise measurements. Every physical quantity that can be quantified is expressed by a relationship of the type $q = nu$, where q is the physical quantity, n is a number, and u is a unit of measurement.

One of the prime concerns of metrologists is to select and define units for all physical quantities.

m.g.f. *See* moment generating function.

Michelson–Morley experiment A famous experiment (1887) to detect the ether, the medium that was supposed to be necessary for the transmission of electromagnetic waves in free space. In the experiment, two light beams were combined to produce interference patterns after travelling for short equal distances perpendicular to each other. The apparatus was then turned through 90° and the two interference patterns were compared to see if there had been a shift of the fringes. If light has a velocity relative to the ether and there is an ether 'wind' as the Earth moves through space, then the times of travel of the two beams would change, resulting in a fringe shift. No shift was detected, not even when the experiment was repeated six months later when the ether wind would have reversed direction. *See also* relativity, theory of.

micro- Symbol: μ A prefix denoting 10^{-6}. For example, 1 micrometre (μm) $= 10^{-6}$ metre (m).

microcomputer *See* computer.

micron Symbol: μm A unit of length equal to 10^{-6} metre.

microprocessor *See* central processor.

mil 1. A unit of length equal to one thousandth of an inch. It is commonly called a 'thou' and is equivalent to 2.54×10^{-5} m. **2.** A unit of area, usually called a *circular mil*, equal to the area of a circle having a diameter of 1 mil.

mile A unit of length equal to 1760 yards. It is equivalent to 1.6093 km.

milli- Symbol: m A prefix denoting 10^{-3}. For example, 1 millimetre (mm) $= 10^{-3}$ metre (m).

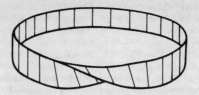

The Möbius strip, which has one
surface and one edge.

million A number equal to 1 000 000 (10^6).

minicomputer *See* computer.

minimum point A point on the graph of a function at which it has the lowest value within an interval. If the function is a smooth continuous curve, the minimum is a turning point, that is, the slope of the tangent to the curve changes continuously from positive to negative by passing through zero. If there is a lower value of the function outside the immediate neighbourhood of the minimum, it is a *local minimum* (or *relative minimum*). If it is lower than all other values of the function it is an *absolute minimum. See also* stationary point, turning point.

minor arc *See* arc.

minor axis *See* ellipse.

minuend The term from which another term is subtracted in a difference. In $5 - 4 = 1$, 5 is the minuend (4 is the subtrahend).

minute (of arc) A unit of plane angle equal to one sixtieth of a degree.

mixed number *See* fraction.

m.k.s. system A system of units based on the metre, the kilogram, and the second. It formed the basis for SI units.

mmHg (millimetre of mercury) A former unit of pressure defined as the pressure that will support a column of mercury one millimeter high under specified conditions. It is equal to 133.322 4 Pa. It is also called the *torr.*

Möbius strip (Möbius band) A continuous flat loop with one twist. It is formed by taking a flat rectangular strip, twisting it in the middle so that each end turns through 180° with respect to the other, and then joining the ends together. Because of the twist, a continuous line can be traced along the surface between any two points, without crossing an edge. The unique topological property of the Möbius strip is that it has one surface and one edge. If a Möbius strip is cut along a line parallel to the edge it is transformed into a single twisted strip that has two edges and two sides. *See also* topology.

modal class The class that occurs with the greatest frequency, for example in a frequency table. *See also* class, mode.

mode The number that occurs most frequently in a set of numbers. For example, the mode (modal value) of $\{5, 6, 2, 3, 2, 1, 2\}$ is 2. If a continuous random variable has probability density function $f(x)$, the mode is the value of x for which $f(x)$ is a maximum. If such a variable has a frequency curve that is approximately symmetrical and has only one mode, then (mean − mode) = 3(mean − median).

modulus The absolute value of a quantity, not considering its sign or direction. For example, the modulus of minus five, written $|-5|$, is 5. The modulus of a vector quantity corresponds to the length or magnitude of the vector. The modulus of a complex number $x + iy$ is $\sqrt{(x^2 + y^2)}$. If the number is written in the form $r(\cos\theta + i\sin\theta)$, the modulus is r. *See also* argument, complex number.

mole Symbol: mol The SI base unit of amount of substance, defined as the amount of substance that contains as many elementary entities as there are atoms in 0.012 kilogram of carbon-12. The elementary entities may be atoms, molecules, ions, electrons, photons, etc., and they must be specified. One mole contains $6.022\,52 \times 10^{23}$ entities. One mole of an element with relative atomic mass A has a mass of A grams (this was formerly called one *gram-atom*). One mole of a compound with relative molecular mass M has a mass of M grams (this was formerly called one *gram-molecule*).

moment (of a force) The turning effect produced by a force about a point. If the point lies on the line of action of the force the moment of the force is zero. Otherwise it is the product of the force and the perpendicular distance from the point to the line of action of the force. If a number of forces are acting on a body, the resultant moment is the algebraic sum of all the individual moments. For a body in equilibrium, the sum of the clockwise moments is equal to the sum of the anticlockwise moments (this law is sometimes called the *law of moments*). *See also* couple, torque.

moment generating function (m.g.f.) A function that is used to calculate the statistical properties of a random variable, x. It is defined in terms of a second variable, t, such that the m.g.f., $M(t)$, is the expectation value of e^{tx}, $E(e^{tx})$. For a discrete random variable
$$M(t) = \Sigma e^{tx}p$$
and for a continuous random variable
$$M(t) = \int e^{tx}f(x)dx$$
Two distributions are the same if their m.g.f.s are the same. The mean and variance of a distribution can be found by differentiating the m.g.f. The mean $E(x) = M'(O)$ and the variance, $\text{Var}(x) = M''(O) - (M'(O))^2$.

moment of area For a given surface, the moment of area is the moment of mass that the surface would have if it had unit mass per unit area.

moment of inertia Symbol: I The rotational analogue of mass. The moment of inertia of an object rotating about an axis is given by
$$I = mr^2$$
where m is the mass of an element a distance r from the axis. *See also* radius of gyration, theorem of parallel axes.

moment of mass The moment of mass of a point mass about a point, line, or plane is the product of the mass and the distance from the point or of the mass and the perpendicular distance from the line or plane. For a system of point masses, the moment of mass is the sum of the mass-distance products for the individual masses. For an object the integral must be used over the volume of the object.

momentum, conservation of *See* constant linear momentum, law of.

momentum, linear Symbol: p The product of an object's mass and its velocity: $p = mv$. The object's momentum cannot change unless a net outside force acts. This relates to Newton's laws and to the definition of force. It also relates to the principle of constant momentum. *See also* angular momentum.

monotonic Always changing in the same direction. A *monotonic increasing function* of a variable x increases or stays constant as x increases, but never decreases. A *monotonic decreasing function* of x decreases or stays constant as x increases, but never increases. Each term in a *monotonic series* is either greater than or equal to the one before it (monotonic increasing) or less than or equal to the one before it (monotonic decreasing). *Compare* alternating series.

multiple A number or expression that has a given number or expression as a factor. For example, 26 is a multiple of 13.

multiple integral *See* iterated integral.

multiple point A point on the curve of a function at which several arcs intersect, or which forms an isolated point, and where a simple derivative of the function does not exist. If the equation of the curve is written in the form:

$$(a_1x + b_1y) + (a_2x^2 + b_2xy + c_2y^2)$$
$$+ (a_3x^3 + \ldots) + \ldots = 0$$

in which the multiple point is at the origin of a Cartesian-coordinate system, the values of the coefficients of x and y indicate the type of multiple point. If a_1 and b_1 are zero, that is, if all the first degree terms are zero, then the origin is a singular point. If the terms a_2, b_2, and c_2 are also zero it is a double point. If, in addition, the terms a_3, b_3, etc., of the third degree terms are zero, it is a *triple point*, and so on. *See also* isolated point, double point.

multiplicand A number or term that is multiplied by another (the *multiplier*) in a multiplication.

multiplication Symbol: × The operation of finding the product of two or more quantities. In arithmetic, multiplication of one number, a, by another, b, consists of adding a to itself b times. This kind of multiplication is commutative, that is, $a \times b = b \times a$. The identity element for arithmetic multiplication is 1, i.e. multiplication by 1 produces no change. In a series of multiplications, the order in which they are carried out does not change the result. For example, $2 \times (4 \times 5) = (2 \times 4) \times 5$. This is the associative law for arithmetic multiplication.

Multiplication of vector quantities and matrices do not follow the same rules.

multiplication of matrices *See* matrix.

multiplication of vectors *See* scalar product, vector product.

multiplier *See* multiplicand.

myria- Symbol: my A prefix used in France to denote 10^4.

N

NAND gate *See* logic gate.

nano- Symbol: n A prefix denoting 10^{-9}. For example, 1 nanometre (nm) $= 10^{-9}$ metre (m).

Napierian logarithm *See* logarithm.

Napier's formulae A set of equations used in spherical trigonometry to calculate the sides and angles in a spherical triangle. In a spherical triangle with sides a, b, and c, and angles opposite these of α, β, and γ respectively:

$$\sin\tfrac{1}{2}(a - b)/\sin\tfrac{1}{2}(a + b) = \tan\tfrac{1}{2}(\alpha - \beta)/\tan\tfrac{1}{2}\gamma$$
$$\cos\tfrac{1}{2}(a - b)/\cos\tfrac{1}{2}(a + b) = \tan\tfrac{1}{2}(\alpha + \beta)/\tan\tfrac{1}{2}\gamma$$
$$\sin\tfrac{1}{2}(\alpha - \beta)/\sin\tfrac{1}{2}(\alpha + \beta) = \tan\tfrac{1}{2}(a - b)/\cot\tfrac{1}{2}c$$
$$\cos\tfrac{1}{2}(\alpha - \beta)/\cos\tfrac{1}{2}(\alpha + \beta) = \tan\tfrac{1}{2}(a + b)/\cot\tfrac{1}{2}c$$

See also spherical triangle.

nappe One of the two parts of a conical surface that lie either side of the vertex. *See* cone.

natural frequency The frequency at which an object or system will vibrate freely. A free vibration occurs when there is no external periodic force and little resistance. The amplitude of free vibrations must not be too great. For instance, a pendulum swinging with small swings under its own weight moves at its natural frequency. Normally, an object's natural frequency is its fundamental frequency.

natural logarithm *See* logarithm.

natural numbers Symbol: N The set of numbers $\{1,2,3, \dots \}$ used for counting separate objects.

necessary condition *See* condition.

negation Symbol: $\sim$ or $\neg$ In logic, the operation of putting *not* or *it is not the*

P	$\sim P$
F	T
T	F

negation

case that in front of a proposition or statement, thus reversing its truth value. The negation of a proposition p is false if p is true and vice versa. The truth-table definition for negation is shown in the illustration. *See also* truth table.

negative Denoting a number of quantity that is less than zero. Negative numbers are also used to denote quantities that are below some specified reference point. For example, in the Celsius temperature scale a temperature of $-24°C$ is 24° below the freezing point of water. *Compare* positive.

negative binomial distribution *See* Pascal's distribution.

neighbourhood *See* topology.

nested intervals A sequence of intervals such that each interval contains the previous one. The *nested interval theorem* states that for any sequence of bounded and closed nested intervals there is at least one point that belongs to all the intervals. If the lengths of the intervals tend to zero as we go through the sequence then there is exactly one such point.

nesting The embedding of a computer subroutine or a loop of instructions within another subroutine or loop, which in turn may lie within yet another, and so on.

net 1. Denoting a weight of goods excluding the weight of the containers or packing.
2. Denoting a profit calculated after deducting all overhead costs, expenses, and taxes.

Compare gross.

neutral equilibrium Equilibrium such that if the system is disturbed a little, there is no tendency for it to move further nor to return. *See* stability.

newton Symbol: N The SI unit of force, equal to the force needed to accelerate one kilogram by one metre per second. 1 N = 1 kg m s^{-2}.

Newtonian mechanics Mechanics based on Newton's laws of motion; i.e. relativistic or quantum effects are not taken into account.

Newton's law of universal gravitation The force of gravitational attraction between two point masses (m_1 and m_2) is proportional to each mass and inversely proportional to the square of the distance (r) between them. The law is often given in the form
$$F = Gm_1m_2/r^2$$
where G is a constant of proportionality called the *gravitational constant*. The law can also be applied to bodies; for example, spherical objects can be assumed to have their mass acting at their centre. *See also* relativity, theory of.

Newton's laws of motion Three laws of mechanics formulated by Sir Isaac Newton in 1687. They can be stated as:
(1) An object continues in a state of rest or constant velocity unless acted on by an external force.
(2) The resultant force acting on an object is proportional to the rate of change of momentum of the object, the change of momentum being in the same direction as the force.
(3) If one object exerts a force on another then there is an equal and opposite force (reaction) on the first object exerted by the second. *See also* reaction.
The first law was discovered by Galileo, and is both a description of inertia and a definition of zero force. The second law provides a definition of force based on the inertial property of mass. The third law is equivalent to the law of conservation of linear momentum.

Newton's method A technique for obtaining successive approximations (iterations) to the solution of an equation, each more accurate than the preceding one. The equation in a variable x is written in the form f(x) = 0, and the general formula or algorithm:
$$x_{n+1} = x_n - f(x_n)/f'(x_n)$$
is applied, where x_n is the nth approximation. Newton's method can be thought of as repeated estimates of the position on a graph of f(x) against x at which the curve crosses the x-axis, by extrapolation of the tangent to the curve. The slope of the tangent at $(x_1,f(x_1))$ is df/dx at $x = x_1$ that is
$$f'(x_1) = f(x_1)/(x_2 - x_1)$$
$x_2 = x_1 - f(x_1)/f'(x_1)$ is therefore the point where the tangent crosses the x-axis, and is a closer approximation to x at f(x) = 0 than x_1 is. Similarly,
$$x_3 = x_2 - f(x_2)/f'(x_2)$$
is a better approximation still. For example, if f(x) = $x^2 - 3$ = 0, then f'(x) = $2x$ and we obtain the algorithm
$$x_{n+1} = x_n - (x^2_n - 3)/2x_n = \tfrac{1}{2}(x_n + 3/x_n)$$
See also iteration.

node A point of minimum vibration in a stationary wave pattern, as near the closed end of a resonating pipe. *Compare* antinode. *See also* stationary wave.

nominal value (per value) The value given to a stock or share by the government or limited company that offers it for sale. Stocks invariably have a nominal value of £100. Shares, however, may have any nominal value. For example, a company wishing to raise £100 000 by an issue of shares may issue 100 000 £1 shares or 200 000 50p shares, or any other combination. The *issue price*, i.e. the price paid by the first buyers of the shares, may not be the same as the nominal value, although it is likely to be close to it. A share with a nominal value of 50p may be offered at an issue price of 55p; it is then said to be offered at a premium of 5p. If offered at an issue price of 45p it is said to be offered at a discount of 5p. Once established as a marketable share on a stock exchange, the nominal value has little importance and it is the *market price* at which it is bought and sold. However, the dividend is always expressed as a percentage of the nominal value.

nomogram A graph that consists of three parallel lines, each one a scale for one of three related variables. A straight line drawn between two points, representing known values of two of the variables, crosses the third line at the corresponding value of the third variable. For example, the lines might show the temperature, volume, and pressure of a known mass of gas. If the volume and pressure are known, the temperature can be read off the nomogram.

nonagon A plane figure with nine straight sides. A *regular nonagon* has nine equal sides and nine equal angles.

non-contradiction, law of *See* laws of thought.

non-Euclidean geometry Any system of geometry in which the parallel postulate of Euclid does not hold. This postulate can be stated in the form that, if a point lies outside a straight line, only one line parallel to the straight line can be drawn through the point. In the early nineteenth century it was shown that it is possible to have a whole self-consistent formal system of geometry without using the parallel postulate at all. There are two types of non-Euclidean geometry. In one (called *elliptic geometry*) there are no parallel lines through the point. An example of this is a system describing the properties of lines, figures, angles, etc., on the surface of a sphere in which all lines are parts of great circles (i.e. circles that have the same centre as the centre of the sphere). Since all great circles intersect, no parallel can be drawn through the point. Note also that the angles of a triangle on such a sphere do not add up to 180°. The other type of non-Euclidean geometry is called a *hyperbolic geometry* – here an infinite number of parallels can be drawn through the point.

Note that a type of geometry is not in itself based on 'experiment' – i.e. of measurements of distance, angles, etc. It is a purely abstract system based on certain assumptions (such as Euclid's axioms). Mathematicians study such systems for their own sake – without necessarily looking for practical applications. The practical applications come in when a particular mathematical system gives an accurate description of physical properties – i.e. the properties of the 'real world'. In practical uses (in architecture, surveying, engineering, etc.) it is assumed that Euclidean geometry applies. However, it is found that this is only an approximation, and that the space-time continuum of relativity theory is non-Euclidean in its properties.

non-isomorphism *See* isomorphism.

NOR gate *See* logic gate.

norm A generalization of the concept of magnitude to any vector space. The norm of a vector x is usually written $\|x\|$. It is a real number associated with the vector and is positive or zero (for the zero vector). If a is a real number, then

$$\|ax\| = a\|x\|$$

and

$$\|x + y\| \leqslant \|x\| + \|y\|$$

normal Denoting a line or plane that is perpendicular to another line or plane. A line or plane is said to be normal to a curve if it is perpendicular to the tangent to the curve at the point at which the line and the curve meet. A radius of a circle, for example, is normal to the circumference. A plane passing through the centre of a sphere is normal to the surface at all points at which they meet.

normal distribution (Gaussian distribution) The type of statistical distribution followed by, for example, the same measurement taken several times, where the variation of a quantity (x) about its mean value (μ) is entirely random. A normal dis-

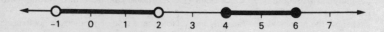

A number line showing an open interval consisting of all the real numbers between – 1 and + 2, and a closed interval from 4 to 6, which includes 4 and 6.

tribution has the probability density function

$$f(x) = \exp[-(x - \mu)^2/2\sigma^2]/\sigma\sqrt{2\pi}$$

where σ is known as the *standard deviation*. The distribution is written $N(\mu,\sigma^2)$. The graph of $f(x)$ is bell-shaped and symmetrical about $x = \mu$. The standard normal distribution has $\mu = 0$ and $\sigma^2 = 1$. x can be standardized by letting $z = (x - \mu)/\sigma$. The values z_α, for which the area under the curve from $-\infty$ to z_α is α, are tabulated; i.e. z is such that $P(z \leqslant z_\alpha) = \alpha$. Hence

$$P(a < x \leqslant b) =$$
$$P(a - \mu)/\sigma < z \leqslant (b - \mu)/\sigma$$

can be found.

normal form *See* canonical form.

normalize To multiply a quantity (e.g. a vector or matrix) by a suitable constant so that its norm is equal to one.

normal subgroup A subgroup H of a group G is *normal* if and only if for any element h in H, $h^{-1}gh$ is in H for all elements g of G.

NOT gate *See* logic gate.

NTP Normal temperature and pressure. *See* STP.

null matrix (zero matrix) A matrix in which all the elements are equal to zero. *See also* matrix.

null set *See* empty set.

number line A straight horizontal line on which each point represents a real number. Integers are points marked at unit distance apart.

numbers Symbols used for counting and measuring. The numbers now in general use are based on the Hindu-Arabic

system, which was introduced to Europe in the 14th and 15th centuries. The Roman numerals used before this made simple arithmetic very difficult, and most calculations needed an abacus. Hindu-Arabic numerals (0, 1, 2, ... 9) enabled calculations to be performed with far greater efficiency because they are grouped systematically in units, tens, hundreds, and so on. *See also* integers, irrational numbers, natural numbers, rational numbers, real numbers, whole numbers.

numerator The top part of a fraction. For example, in the fraction $\frac{3}{4}$, 3 is the numerator and 4 is the denominator. The numerator is the dividend.

numerical analysis The study of methods of calculation that involve approximations, for example, iterative methods. *See also* iteration.

numerical integration A procedure for calculating approximate values of integrals. Sometimes a function is known only as a set of values for corresponding values of a variable and not as a general formula that can be integrated. Also, many functions cannot be integrated in terms of known standard integrals. In these cases, numerical integration methods, such as the trapezium rule and Simpson's rule, can be used to calculate the area under a graph corresponding to the integral. The area is divided into vertical columns of equal width, the width of each column representing an interval between two values of x for which $f(x)$ is known. Usually a calculation is first carried out with a few columns; these are further subdivided until the desired accuracy is attained, i.e. when further subdivision makes no significant difference to the result. *See also* Simpson's rule, trapezium rule.

O

object The set of points that undergoes a geometrical transformation or mapping. *See also* projection.

oblate Denoting a spheroid that has a polar diameter that is smaller than the equatorial diameter. The Earth, for example, is not a perfect sphere but is an oblate spheroid. *Compare* prolate. *See also* ellipsoid.

oblique Forming an angle that is not a right angle.

oblique coordinates *See* Cartesian coordinates.

oblique solid A solid geometrical figure that is 'slanted'; for example, a cone, cylinder, pyramid, or prism with an axis that is not at right angles to its base. *Compare* right solid.

oblique spherical triangle *See* spherical triangle.

oblique triangle A triangle that does not contain a right angle.

obtuse Denoting an angle that is greater than 90° but less than 180°. *Compare* acute.

OCR (optical character recognition) A system used to input information to a computer. The information, usually in the form of letters and numbers, is printed, typed, or sometimes hand-written. The characters used can be read and identified optically by an *OCR reader*. This machine interprets each character and translates it into a series of electrical pulses.

octagon A plane figure with eight straight sides. A *regular octagon* has eight equal sides and eight equal angles.

octahedron A polyhedron that has eight faces. A *regular octahedron* has eight faces, each one an equilateral triangle. *See also* polyhedron.

octal Denoting or based on the number eight. An octal number system has eight different digits instead of the ten in the decimal system. Eight is written as 10, nine as 11, and so on. *Compare* binary, decimal, duodecimal, hexadecimal.

octant 1. One of eight regions into which space is divided by the three axes of a three-dimensional Cartesian coordinate system. The first octant is the one in which x, y, and z are all positive. The second, third, and fourth octants are numbered anticlockwise around the positive z-axis. The fifth octant is underneath the first, the sixth under the second, etc.
2. A unit of plane angle equal to 45 degrees ($\pi/4$ radians).

odd Not divisible by two. The set of odd numbers is {1, 3, 5, 7, ... }. *Compare* even.

odd function A function $f(x)$ of a variable x for which $f(-x) = -f(x)$. For example, $\sin x$ and x^3 are odd functions of x. *Compare* even function.

odds When bets are placed on some event the odds are the probability of it happening.

oersted Symbol Oe A unit of magnetic field strength in the c.g.s. system. It is equal to $10^3/4\pi$ amperes per metre ($10^3/4\pi$ A m^{-1}).

off-line Disconnected from or not under the direct control of a computer. Off-line equipment is either not in use, is undergoing repair, or is performing some task without the assistance of the central processor of the computer. *Compare* online.

ohm Symbol: Ω The SI unit of electrical resistance, equal to a resistance that passes a current of one ampere when there is an electric potential difference of one volt across it. $1\ \Omega = 1\ V\,A^{-1}$. Formerly, it was defined in terms of the resistance of a column of mercury under specified conditions.

one-to-one correspondence A function or mapping between two sets of things or numbers, such that each element in the first set maps into only one element in the second, and vice versa. *See also* function, homomorphism.

on-line Connected to and controlled by a computer. Any device that is connected to and capable of interacting directly with a computer without human intervention is said to be on-line. In *on-line processing* the processing of a computer program is performed on equipment directly controlled by the central processor. *Compare* off-line.

onto A mapping from one set S to another set T is *onto* if every member of T is the image of some member of S under the mapping. *Compare* into.

open curve A curve in which the ends do not meet, for example, a parabola or a hyperbola. *Compare* closed curve.

open interval A set consisting of the numbers between two given numbers (end points), not including the end points, for example, all the real numbers greater than 1 and less than 4.5 constitute an open interval. The open interval between two real numbers a and b is written (a,b). Here, the round brackets indicate that the points a and b are not included in the interval. On a number line, the end points of an open interval are circled. *Compare* closed interval. *See also* interval.

open sentence In formal logic, a sentence that contains one or more free variables.

open set A set defined by limits that are not included in the set itself. The set of all rational numbers greater than 0 and less than ten, written $\{x : 0 < x < 10; x \in R\}$, and the set of all the points inside a circle, but not including the circle itself, are examples of open sets. *Compare* closed set.

operating system (OS) The collection of programs used in the control of a computer system. It is generally supplied by the computer manufacturer. An operating system has to decide at any instant which of the many demands on the attention of the central processor to satisfy next. These demands include input from and output to various devices, the execution of a number of programs, and accounting and timing. Large computers, in which many jobs can be run simultaneously, will have a highly complex operating system. A program that runs without the benefit of an operating system is called a *stand-alone program*.

operator 1. A mathematical function, such as addition, subtraction, multiplication, or taking a square root or a logarithm, etc. *See* function.
2. The symbol denoting a mathematical operation or function, for example: $+$, $-$, $\times$, $\sqrt{}$, $\log_{10}$.

opposite Denoting the side facing a given angle in a triangle, i.e. the side not forming one arm of the angle. In trigonometry, the ratios of the length of the opposite side to the other side lengths in a right-angled triangle are used to define the sine and tangent functions of the angle.

optical character recognition *See* OCR.

or *See* disjunction.

orbit The curved path or trajectory along which a moving object travels under the influence of a gravitational field. An object with a negligible mass moving under the influence of a planet or other body has an orbit that is a conic section; i.e. a parabola, ellipse, or hyperbola.

order 1. (of a matrix) The number of rows and columns in a matrix. *See* matrix.
2. (of a derivative) The number of times a variable is differentiated. For example, dy/dx is a first-order derivative; d^2y/dx^2 is second-order; etc.
3. (of a differential equation) The order of the highest derivative in an equation. For example,

$$d^3y/dx^3 + 4xd^2y/dx^2 = 0$$

is a third-order differential equation.

$$d^2y/dx^2 - 3x(dy/dx)^3 = 0$$

is a second-order differential equation. *Compare* degree. *See also* differential equation.

ordered pair Two numbers indicating values of two variables in a particular order. For example, the x- and y-coordinates of points in a two-dimensional Cartesian coordinate system form a set of ordered pairs (x,y).

ordered set A set of entities in a particular order. *See* sequence.

ordered triple Three numbers indicating values of three variables in a particular order. The x-, y-, and z-coordinates of a point in a three-dimensional coordinate system form an ordered triple (x,y,z).

ordinal numbers Whole numbers that denote order, as distinct from number or quantity. That is, first, second, third, and so on. *Compare* cardinal numbers.

ordinary differential equation An equation that contains total derivatives but no partial derivatives. *See* differential equation.

ordinate The vertical coordinate (y-coordinate) in a two-dimensional rectangular Cartesian coordinate system. *See* Cartesian coordinates.

OR gate *See* logic gate.

origin The fixed reference point in a coordinate system, at which the values of all the coordinates are zero and at which the axes meet. *See* coordinates.

orthocentre A point in a triangle that is the point of intersection of lines from each vertex perpendicular to the opposite sides. The triangle formed by joining the feet of these vertices is the *pedal triangle*.

orthogonal projection A geometrical transformation that produces an image on a line or plane by perpendicular lines crossing the plane. If a line of length l is projected orthogonally from a plane at angle θ to the image plane, its image length is $l\cos\theta$. The image of a circle is an ellipse. *See also* projection.

OS *See* operating system.

oscillating series A special type of nonconvergent series for which the sum does not approach a limit but continually fluctuates. Oscillating series can either fluctuate between bounds, for example the series $1 - 1 + 1 - 1 + \ldots$, or it can be unbounded, for example $1 - 2 + 3 - 4 + 5 - \ldots$.

oscillation A regularly repeated motion or change. *See* vibration.

ounce 1. A unit of mass equal to one sixteenth of a pound. It is equivalent to 0.028349 kg.
2. A unit of capacity, often called a *fluid ounce*, equal to one twentieth of a pint. In the UK, it is equivalent to 2.8413×10^{-5} m^3. In the USA a fluid ounce is equal to one sixteenth of a US pint. It is equivalent to 2.0573×10^{-5} m^3. 1 UK fluid ounce is equal to 0.9608 US fluid ounce.

output 1. The signal or other form of information obtained from an electrical device, machine, etc. The output of a computer is the information or results derived from the data and programmed instructions fed into it. This information is transferred as a series of electrical pulses from the central processor of the computer to a selected *output device*. Some of these output units convert the pulses into a readable or pictorial form; examples include the printer, plotter, and visual display unit (which can also be used as an input device). Other output devices translate the pulses into a form that can be fed back into the computer at a later stage; the magnetic tape unit is an example.

2. The process or means by which output is obtained.

3. To deliver as output.

See also input, input/output.

overdamping *See* damping.

overflow The situation arising in computing when a number, such as the result of an arithmetical operation, has a greater magnitude than can be represented in the space allocated to it in a register or a location in store.

overlay A technique used in computing when the total storage requirements for a lengthy program exceeds the space available in the main store. The program is split into sections so that only the section or sections required at any one time will be transferred into the main store from a disk unit or other backing store. These program segments (or *overlays*) will all occupy the same area of the main store. The overlay structure must be organized so that no routine calls another that would overwrite it.

P

paper tape A long strip of paper, or sometimes thin flexible plastic, on which information can be recorded as a pattern of round holes punched in rows across the tape. The positions at which holes may be punched are called *tracks*; one-inch tape with eight tracks per row is widely used. There is also a line of small sprocket holes along the length of the tape between tracks three and four. A digit (0–9), letter, or some other character is represented on the tape by a particular combination of holes in a row; when eight tracks are used for representing characters, there are 2^8 or 256 possible combinations of holes and hence 256 characters can be represented. Paper tape has been used to input and output information in a wide variety of devices. The punched information is fed into a computer using a *paper-tape reader*. This machine senses the presence or absence of holes in each row and converts the information into a series of electric pulses. (A hole usually produces a pulse, a non-hole produces no pulse.) The pulses are then transmitted to the central processor of the computer. Although maybe 1000 rows can be read per second, the paper-tape reader is a slow input device. Information is output on paper tape using a *paper-tape punch*, which automatically punches data into the tape. Punched tape from a computer can be fed back in at a later date or fed into another computer. *Compare* card, magnetic tape, disk.

Pappus' theorems Two theorems concerning the rotation of a curve or a plane shape about a line that lies in the same plane. The first theorem states that the surface area generated by a curve revolving about a line that does not cross it, is equal to the length of the curve times the circumference of the circle traced out by its centroid. The second theorem states that the volume of a solid of revolution generated by a plane area that rotates about a line not crossing it, is equal to the area times the circumference of the circle traced out by the centroid of the area. (Note that the plane area and the line both lie in the same plane.)

parabola A conic with an eccentricity of 1. The curve is symmetrical about an axis through the focus at right angles to the directrix. This axis intercepts the parabola at the *vertex*. A chord through the focus perpendicular to the axis is the *latus rectum* of the parabola.
In Cartesian coordinates a parabola can be represented by an equation:
$$y^2 = 4ax$$
In this form the vertex is at the origin and the x-axis is the axis of symmetry. The focus is at the point $(0,a)$ and the directrix is the line $x = -a$ (parallel to the y-axis). The *latus rectum* is $4a$.
If a point is taken on a parabola and two lines drawn from it – one parallel to the axis and the other from the point to the focus – then these lines make equal angles with the tangent at that point. This is known as the *reflection property* of the parabola, and is utilized in parabolic reflectors and antennae. *See* paraboloid.
The parabola is the curve traced out by a projectile falling freely under gravity. For example, a tennis ball projected horizontally with a velocity v has, after time t, travelled a distance $d = vt$ horizontally, and has also fallen vertically by $h = gt^2/2$ because of the acceleration of free fall g. These two equations are *parametric equations* of the parabola. Their standard form, corresponding to $y^2 = 4ax$, is
$$x = at^2$$
$$y = 2at$$
where x represents h, the constant a is $g/2$, and y represents d. *See also* conic.

paraboloid A curved surface in which the cross-sections in any plane passing through a central axis is a parabola. A *paraboloid of revolution* is formed when a parabola is rotated around its axis of symmetry. Parabolic surfaces are used in telescope mirrors, searchlights, radiant

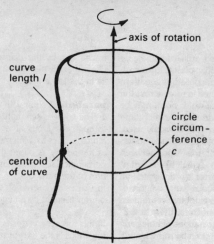

Pappus' theorem: the curved surface
area $A = l \times c$

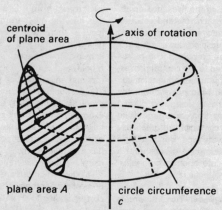

Pappus' theorem: the volume
enclosed by the curved surface
$V = A \times c$

heaters, and radio antennae on account of
the focusing property of the parabola.
Another type of paraboloid is the *hyperbolic paraboloid*. This is a surface with the
equation:

$$x^2/a^2 - y^2/b^2 = 2cz$$

where c is a positive constant. Sections
parallel to the xy plane ($z = 0$) are
hyperbolas. Sections parallel to the other
two planes ($x = 0$ or $y = 0$) are parabo-
las.

paradox (antinomy) A proposition or statement that leads to a contradiction if it is asserted *and* if it is denied.

A famous example of a paradox is *Russell's paradox* in set theory. A set is a collection of things. It is possible to think of sets as belonging to two groups: sets that contain themselves and sets that do not contain themselves. A set that contains itself is itself a member of the set. For instance, the set of all sets is itself a set, so it contains itself. Other examples are the 'set of all things that one can think about' and 'the set of all abstract ideas'. On the other hand the set of all things with four legs does not itself have four legs – it is an example of a set that does not contain itself. Other examples of sets that do not contain themselves (or are not members of themselves) are 'the set of all oranges' and 'the set of all things that are green'. Consider now the set of all sets that do not contain themselves. The paradox arises from the question, 'Does the set of all sets that do not contain themselves contain itself or not?' If one asserts that it does contain itself then it cannot be one of the things that do not contain themselves. This means that it does not belong to the set, so it doesn't contain itself. On the other hand, if it does not contain itself it must be one of all the things that do not contain themselves, so it must belong to the set. The short answer to the question is, 'If it does, it doesn't; if it doesn't, it does!' – hence the paradox. Paradoxes like this can be used in investigating fundamental questions in set theory.

parallax The angle between the direction of an object, for example a star or a planet, from a point on the surface of the Earth and the direction of the same object from the centre of the Earth. The measurement of parallax is used to find the distance of an object from the Earth. The object is viewed from two widely separated points on the Earth and the distance between these two points and the direction of the object as seen from each is measured. The parallax at the observation points and the distance of the object from the observer can be found by simple trigonometry.

parallel Extending in the same direction and remaining the same distance apart. *Compare* antiparallel.

parallel axes, theorem of *See* theorem of parallel axes.

parallel forces When the forces on an object pass through one point, their resultant can be found by using the parallelogram of vectors. If the forces are parallel the resultant is found by addition, taking sign into account. There may also be a turning effect in such cases, which can be found by the principle of moments.

parallelogram A plane figure with four straight sides, and with opposite sides parallel and of equal length. The opposite angles of a parallelogram are also equal. Its area is the product of the length of one side and the perpendicular distance from that side to the side opposite. In the special case in which the angles are all right angles the parallelogram is a rectangle; when all the sides are equal it is a rhombus.

parallelogram (law) of forces *See* parallelogram of vectors.

parallelogram (law) of velocities *See* parallelogram of vectors.

parallelogram of vectors A method for finding the resultant of two vectors acting at a point. The two vectors are shown as two adjacent sides of a parallelogram: the resultant is the diagonal through the starting point. The technique can be used either with careful scale drawing or with trigonometry. The trigonometrical relations give:
$$F = \sqrt{(F_1^2 + F_2^2 + 2F_1F_2\cos\theta)}$$
$$\alpha = \sin^{-1}[(F_2/F)\sin\theta]$$
where θ is the angle between F_1 and F_2 and α is the angle between F and F_1. *See* vector.

Pascal's triangle

It need not be the case that $x < y$ or $y < x$ for any two elements x and y. An example of a partially ordered set is the set of subsets of a given set where we define $A < B$ for sets A and B to mean that A is a proper subset of B.

partial sum The sum of a finite number of terms from the start of an infinite series. In a convergent series, the partial sum of the first r terms, S_r, is an approximation to the sum to infinity. *See* series.

particle An abstract simplification of a real object – the mass is concentrated at the object's centre of mass; its volume is zero. Thus rotational aspects can be ignored.

partition A partition of a set S is a finite collection of disjoint sets whose union is S.

pascal Symbol: Pa The SI unit of pressure, equal to a pressure of one newton per square metre ($1 \text{ Pa} = 1 \text{ N m}^{-2}$). The pascal is also the unit of stress.

Pascal's distribution (negative binomial distribution) The distribution of the number of independent Bernoulli trials performed up to and including the rth success. The probability that the number of trials, x, is equal to k is given by
$$P(x=k) = {}^{k-1}C_{r-1}p^rq^{k-r}$$
The mean and variance are r/p and rq/p^2 respectively. *See also* geometric distribution.

Pascal's triangle A triangle array of numbers in which each row starts and ends with 1 and that is built up by summing two adjacent numbers in a row to obtain the number directly below them in the next row. Each row in Pascal's triangle is set of binomial coefficients. In the expansion of $(x + y)^n$, the coefficients of x and y are given by the $(n + 1)^{\text{th}}$ row of Pascal's triangle.

pedal curve The *pedal curve* of a given curve C with respect to a fixed point P is the locus of the foot of the perpendicular from P to a variable tangent to the curve C. The point P is called the *pedal point*. For example, if C is a circle and P is a point on its circumference the pedal curve is a cardioid.

pedal triangle *See* orthocentre.

pencil A family of geometric objects that share a common property. For example, a pencil of circles consists of all the circles in a given plane that pass through two given points, a pencil of lines consists of all the lines in a given plane passing through a given point, and a pencil of parallel lines consists of all the lines parallel to a given direction.

pendulum A body that oscillates freely under the influence of gravity. A *simple pendulum* consists of a small mass oscillating to and fro at the end of a very light string. If the amplitude of oscillation is small (less than about $10°$), it moves with simple harmonic motion; the period does not depend on amplitude. There is a continuous interchange of potential and kinetic energy through the motion; at the ends of the swings the potential energy is a maximum and the kinetic energy zero. At the mid-point the kinetic energy is a maximum and the potential energy is zero. The period is given by

$$T = 2\pi\sqrt{(l/g)}$$

Here l is the length of the pendulum (from support to centre of the mass) and g is the acceleration of free fall.

A *compound pendulum* is a rigid body swinging about a point. The period of a compound pendulum depends on the moment of inertia of the body. For small oscillations it is given by the same relationship as that of the simple pendulum with l replaced by $[\sqrt{(k^2 + h^2)}]/h$. Here, k is the radius of gyration about an axis through the centre of mass and h is the distance from the pivot to the centre of mass.

pentagon A plane figure with five straight sides. In a *regular pentagon*, one with all five sides and angles equal, the angles are all 108°. A regular pentagon can be superimposed on itself after rotation through 72° ($2\pi/5$ radians).

percentage A number expressed as a fraction of one hundred. For example, 5 percent (or 5%) is equal to 5/100. Any fraction or decimal can be expressed as a percentage by multiplying it by 100. For example, $0.63 \times 100 = 63\%$ and $\frac{1}{4} \times 100 = 25\%$.

percentage error The error or uncertainty in a measurement. For example, if, in measuring a length of 20 metres, a tape can measure to the nearest four centimetres, the measurement is written as $20\pm$ 0.04 metres and the percentage error is $(0.04/20) \times 100 = 0.2\%$. *See also* error.

percentile One of the set of points that divide a set of data arranged in numerical order into 100 parts. The rth percentile, P_r, is the value below and including which $r\%$ of the data lies and above which $(100 - r)\%$ lies. P_r can be found from the cumula-

tive frequency graph. *See also* quartile, range.

perfect number A number that is equal to the sum of all its factors except itself. 28 is a perfect number since its factors are 1, 2, 4, 7, and 14, and $1 + 2 + 4 + 7 + 14 = 28$.

perimeter The distance round the edge of a plane figure. For example, the perimeter of a rectangle is twice the length plus twice the breadth. The perimeter of a circle is its circumference ($2\pi r$).

period Symbol: T The time for one complete cycle of an oscillation, wave motion, or other regularly repeated process. It is the reciprocal of the frequency, and is related to pulsatance, or angular frequency, (ω) by $T = 2\pi/\omega$.

periodic function A function that repeats itself at regular intervals of the variable. For example $\sin x$ is a periodic function of x because $\sin x = \sin(x + 2\pi)$ for all values of x.

periodic motion Any kind of regularly repeated motion, such as the swinging of a pendulum, the orbiting of a satellite, the vibration of a source of sound, or an electromagnetic wave. If the motion can be represented as a pure sine wave, it is a simple harmonic motion. Harmonic motions in general are given by the sum of two or more pure sine waves.

period of investment The length of time for which a fixed amount of capital remains invested. In times of historically low interest rates, an investor prepared to commit his money for a long period, such as five or ten years, will gain a higher rate of interest than he can expect for a short-term investment. However, if interest rates are historically high this will not be the case and long-term rates may be lower than short-term rates.

peripheral unit (peripheral) A device connected to and controlled by the central processor of a computer. Peripherals include input devices, output devices, and backing store. Some examples are visual display units, printers, magnetic tape units, and disk units. *See also* input, output.

permutation An ordered subset of a given set of objects. For three objects, A, B, and C, there are six possible permutations: ABC, ACB, BAC, BCA, CAB, and CBA. The total number of permutations of n objects is n!
The total number of permutations of r objects taken from n objects is given by $n!/(n - r)!$, assuming that each object can be selected only once. This is written ^nP_r. For example, the possible permutations of two objects from the set of three objects A, B, and C would be AB, BA, AC, CA, BC, CB. Note that each object is selected only once in this case – if the objects could occur any number of times, the above set of permutations would include AA, BB, and CC. The number of permutations of r objects selected from n objects when each can occur any number of times is n^r.
Note also the difference between permutations and combinations: permutations are different if the order of selection is different, so AB and BA are different permutations but the same combination. The number of combinations of r objects from n objects is written ^nC_r, and $^nP_r = {}^nC_r \times r!$

perpendicular At right angles. The perpendicular bisector of a line crosses it half way along its length and forms a right angle. A vertical surface is perpendicular to a horizontal surface.

phase The stage in a cycle that a wave (or other periodic system) has reached at a particular time (taken from some reference point). Two waves are *in phase* if their maxima and minima coincide.
For a simple wave represented by the equation
$$y = a\sin 2\pi(ft - x/\lambda)$$
The phase of the wave is the expression
$$2\pi(ft - x/\lambda)$$
The *phase difference* between two points distances x_1 and x_2 from the origin is
$$2\pi(x_1 - x_2)/\lambda$$

A more general equation for a progressive wave is
$$y = a\sin 2\pi[ft - (x/\lambda) - \phi]$$
Here, ϕ is the *phase constant* – the phase when t and x are zero. Two waves that are out of phase have different stages at the origin. The phase difference is $\phi_1 - \phi_2$. It is equal to $2\pi x/\lambda$, where x is the distance between corresponding points on the two waves. It is the *phase angle* between the two waves; the angle between two rotating vectors (phasors) representing the waves. *See also* wave.

phase angle *See* phase.

phase constant *See* phase.

phase difference *See* phase.

phase speed The speed with which the phase in a travelling wave is propagated. It is equal to λ/T, where T is the period. *Compare* group speed.

phasor *See* simple harmonic motion.

pi (π) The ratio of the circumference of any circle to its diameter. π is approximately equal to $3.14159\ldots$ and is a transcendental number (its exact value cannot be written down, but it can be stated to any degree of accuracy).

pico- Symbol: p A prefix denoting 10^{-12}. For example, 1 picofarad (pF) = 10^{-12} farad (F).

pictogram (pictograph) A diagram that represents statistical data in a pictorial form. For example, the proportions of pink, red, yellow, and white flowers that grow from a packet of mixed seeds can be shown by rows of the appropriate relative numbers of coloured flower shapes.

piecewise A function is *piecewise continuous* on S if it is defined on S and can be separated into a finite number of pieces such that the function is continuous on the interior of each piece. Terms such as *piecewise differentiable* and *piecewise linear* are similarly defined.

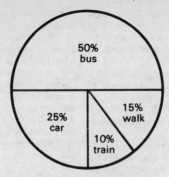

A pie chart showing how a group of workers travel to work.

pie chart A diagram in which proportions are illustrated as sectors of a circle, the relative areas of the sectors representing the different proportions. For example, if, out of 100 workers in a factory, 25 travel to work by car, 50 by bus, 10 by train, and the rest walk, the bus passengers are represented by half of the circle, the car passengers by a quarter, the train users by a 36° sector, and so on.

pint A unit of capacity. In the UK it is equal to one eight of a UK gallon and is equivalent to $5.682\,6 \times 10^{-4}$ m³. The US liquid pint is equal to one eighth of a US gallon and is equivalent to $4.731\,8 \times 10^{-4}$ m³. The US dry pint is equal to one sixty-fourth of a US bushel and is equivalent to $5.506\,1 \times 10^{-4}$ m³.

pixel *See* computer graphics.

PL/1 *See* program.

plane A flat surface, either real or imaginary, in which any two points are joined by a straight line lying entirely on the surface. *Plane geometry* involves the relationships between points, lines, and curves lying in the same plane. In Cartesian coordinates, any point in a plane can be defined by two coordinates, x and y. In three-dimensional coordinates, each value of z corresponds to a plane parallel to the plane in which the

x and y axes lie. For any three points, there exists only one plane containing all three. A particular plane can also be specified by a straight line and a point.

plot To draw on a graph. A series of individual points plotted on a graph may show a general relationship between the variables represented by the horizontal and vertical axes. For example, in a scientific experiment one quantity can be represented by x and another by y. The values of y at different values of x are then plotted as a series of points on a graph. If these fall on a line or curve, then the line or curve drawn through the points is said to be a plot of y against x.

plotter An output device of a computer system that produces a permanent record of the results of some program by drawing lines on paper. One pen, or maybe two or more pens with different coloured ink, are moved over the paper according to instructions sent from the computer or from a backing store. Plotters are used for drawing graphs, contour maps, etc.

plotting The process of marking points on a system of coordinates, or of drawing a graph by marking points.

point A location in space, on a surface, or in a coordinate system. A point has no

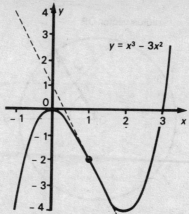

The graph of $y = x^3 - 3x^2$ has a point of inflection at $x = 1$, $y = -2$. The derivative $dy/dx = -3$ at this point.

dimensions and is defined only by its position.

point of contact A single point at which two curves, or two curved surfaces touch. There is only one point of contact between the circumference of a circle and tangent to the circle. Two spheres also can have only one point of contact.

point of inflection A point on a curved line at which the tangent changes its direction of rotation. Approaching from one side of the point of inflection, the slope of the tangent to the curve increases; and moving away from it on the other side, it decreases. For example, the graph of $y = x^3 - 3x^2$ in rectangular Cartesian coordinates, has a point of inflection at the point $x = 1$, $y = -2$. The second derivative d^2y/dx^2 on the graph of a function $y = f(x)$ is zero and changes its sign at a point of inflection. Thus, in the example above, $d^2y/dx^2 = 6x - 6$, which is equal to zero at the point $x = 1$.

Poisson distribution A probability distribution for a discrete random variable. It is defined, for a variable (r) that can take

values in the range $0, 1, 2, \ldots$, and has a mean value μ, as
$$P(r) = e^- m^r/r!$$
A binomial distribution with a small frequency of success p in a large number n of trials can be approximated by a Poisson distribution with mean np.

polar coordinates A method of defining the position of a point by its distance and direction from a fixed reference point (pole). The direction is given as the angle between the line from the origin to the point, and a fixed line (axis). On a flat surface only one angle, θ, and the radius, r, are needed to specify each point. For example, if the axis is horizontal, the point $(r,\theta) = (1,\pi/2)$ is the point one unit length away from the origin in the perpendicular direction. Conventionally, angles are taken as positive in the anticlockwise sense. In a rectangular Cartesian coordinate system with the same origin and the x-axis at $\theta = 0$, the x- and y-coordinates of the point (r,θ) are
$$x = r\cos\theta$$
$$y = r\sin\theta$$
Conversely
$$r = \sqrt{(x^2 + y^2)}$$

141

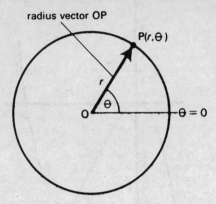

radius vector OP

P(r,Θ)

Θ = 0

The point P(r,Θ) in two-dimensional polar coordinates.

and
$$\theta = \tan^{-1}(y/x)$$
In three dimensions, two forms of polar coordinate systems can be used. *See* cylindrical polar coordinates, spherical polar coordinates. *See also* Cartesian coordinates.

pole 1. One of the points on the Earth's surface through which its axis of rotation passes, or the corresponding point on any other sphere.
2. *See* stereographic projection.
3. *See* polar coordinates.

Polish notation A notation in which parentheses are unnecessary since all formulae can be written unambiguously without them. In Polish notation operators precede their operands. Thus, $a + b$ is written $+ab$. The notation was invented by a Polish mathematician, Jan Lukasiewicz.

polygon A plane figure bounded by a number of straight sides. In a *regular polygon*, all the sides are equal and all the internal angles are equal. In a regular polygon of n sides the exterior angle is $360°/n$.

polyhedron A solid figure bounded by plane polygonal faces. The point at which three or more faces intersect on a polyhedron is called a *vertex*, and a line along which two faces intersect is called an *edge*. In a *regular polyhedron*, all the faces are congruent regular polygons. There are only five regular polyhedrons: the regular tetrahedron, which has four equilateral triangular faces; the regular hexahedron, or cube, which has six equilateral square faces; the regular octahedron, which has eight equilateral triangular faces; the regular dodecahedron, which has twelve regular pentagonal faces; and the regular icosahedron, which has twenty equilateral triangular faces. These are all *convex polyhedrons*. That is, all the angles between faces and edges are convex and the polyhedron can be laid down flat on any one of the faces. In a *concave polyhedron*, there is at least one face in a plane that cuts through the polyhedron. The polyhedron cannot be laid down on this face.

polynomial A sum of multiples of integer powers of a variable. The general equation for a polynomial in the variable x is
$$a_0x^n + a_1x^{n-1} + a_2x^{n-2} + \ldots$$

where a_0, a_1, etc., are constants and n is the highest power of x, called the *degree* of the polynomial. If $n = 1$, it is *linear* expression, for example, $f(x) = 2x + 3$. If $n = 2$, it is *quadratic*, for example, $x^2 + 2x + 4$. If $n = 3$, it is *cubic*, for example, $x^3 + 8x^2 + 2x + 2$. If $n = 4$, it is *quartic*. If $n = 5$, it is *quintic*.

On a Cartesian coordinate graph on which $(n + 1)$ individual points are plotted, there is at least one polynomial curve that passes through all the points. By choosing suitable values of a_0 and a_1, the straight line

$$y = a_0 x + a_1$$

can be made to pass through any two points. Similarly a quadratic

$$y = a_0 x^2 + a_1 x + a_2$$

can be made to pass through any three points.

A polynomial may have more than one variable:

$$4x^2 + 2xy + y^2$$

is a polynomial of degree 2 (second-degree polynomial) of two variables.

polytope The analogue in n dimensions of point, line, polygon, and polyhedron in 0, 1, 2, and 3 dimensions respectively.

position vector The vector that represents the displacement of a point from a given reference origin. If a point P in polar coordinates has coordinates (r, θ), then $\mathbf{r}$ is the position vector of P – a vector of magnitude r making an angle θ with the axis. *See* vector.

positive Denoting a number or quantity that is greater than zero. Numbers that are used in counting things and measuring sizes are all positive numbers. If a change in a quantity is positive, it increases, that is it moves away from zero if it is already positive, and towards it if it is negative. *Compare* negative.

postulate *See* axiom.

potential energy Symbol: V The work an object can do because of its position or state. There are many examples. The work an object at height can do in falling is

its gravitational potential energy. The energy 'stored' in elastic or a spring under tension or compression is elastic potential energy. Potential difference in electricity is a similar concept, and so on. In practice the potential energy of a system is the energy involved in bringing it to its current state from some reference state; i.e. it is the same as the work that the system could do in moving from its current state back to a reference state. *See also* energy.

pound A unit of mass now defined as 0.453 592 37 kg.

poundal Symbol: pdl The unit of force in the f.p.s. system. It is equal to 0.138 255 newton (0.138 255 N).

power 1. The number of times a quantity is to be multiplied by itself. For example, $2^4 = 2 \times 2 \times 2 \times 2 = 16$ is known as the fourth power of two, or two to the power four. *See also* exponent, power series. **2.** Symbol: P The rate of energy transfer (or work done) by or to a system. The SI unit of power is the watt – the energy transfer in joules per second.

power series A series in which the terms contain regularly increasing powers of a variable. For example,

$$S_n = 1 + 2x + 3x^2 + 4x^3 + \ldots + nx^{n-1}$$

is a power series in the variable x. In general, a power series has the form

$$a_0 + a_1 x + a_2 x^2 + \ldots + a_n x^n$$

where a_0, a_1, etc. are constants.

precession If an object is spinning on an axis and a force is applied at right angles to this axis, then the axis of rotation can itself move around another axis at an angle to it. The effect is seen in tops and gyroscopes, which 'wobble' slowly while they spin as a result of the force of gravity. The Earth also precesses – the axis of rotation slowly describes a cone. The precession of Mercury is a movement of the whole orbit of the planet around an axis perpendicular to the orbital plane. It can be explained by relativistic mechanics.

precision The number of figures in a number. For example 2.342 is stated to a precision of four significant figures, or three decimal places. The precision of a number normally reflects the accuracy of the value it represents. *See also* accuracy.

premiss In logic, an initial proposition or statement that is known or assumed to be true and on which a logical argument is based. *See* logic.

premium 1. The difference between the issue price of a stock or share and its nominal value when the issue price is in excess of the nominal value. *Compare* discount. **2.** The amount of money paid each year to an insurance company to purchase insurance cover for a specified risk.

pressure Symbol: *p* The pressure on a surface due to forces from another surface or from a fluid is the force acting at right angles to unit area of the surface:
pressure = force/area.
The unit is the pascal (Pa), equal to the newton per square metre.
Objects are often designed to maximize or minimize pressure applied. To give maximum pressure, a small contact area is needed – as with drawing pins and knives. To give minimum pressure, a large contact area is needed – as with snow shoes and the large tyres of certain vehicles.
Where the pressure on a surface is caused by the particles of a fluid (liquid or gas), it is not always easy to find the force on unit area. The pressure at a given depth in a fluid is the product of the depth, the average fluid density, and *g* (the acceleration of free fall):
pressure in a fluid = depth × mean density × *g*
As it is normally possible to measure the mean density of a liquid only, this relation is usually restricted to liquids.
The pressure at a point at a certain depth in a fluid:
(1) is the same in all directions;
(2) applies force at 90° to any contact surface;
(3) does not depend on the shape of the container.

pressure of the atmosphere The pressure at a point near the Earth's surface due to the weight of air above that point. Its value varies around about 100 kPa (100 000 newtons per square metre).

prime A *prime number* is a positive integer which is not 1 and has no factors except 1 and itself. The set of prime numbers is {2, 3, 5, 7, 11, 13, 17, 19, 23, 29, . . . }. There are an infinite number of prime numbers but no general formula for them. The *prime factors* of a number are the prime numbers that divide into it exactly. For example, the prime factors of 45 are 3 and 5 since 45 = 3 × 3 × 5. Each whole number has a unique set of prime factors. *See* Eratosthenes, sieve of.

principal A sum of money that is borrowed, on which interest is charged. *See* compound interest, simple interest.

principal diagonal *See* square matrix.

principle of equivalence *See* relativity, theory of.

principle of moments The principle that when an object or system is in equilibrium the sum of the moments in any direction equals the sum of the moments in the opposite direction. Because there is no resultant turning force, the moments of the forces can be measured relative to any point in the system or outside it.

printout The computer output, in the form of characters printed on a continuous sheet of paper, produced by a line printer or similar device. *See* output, line printer.

prism A polyhedron with two parallel opposite faces, called *bases*, that are congruent polygons. All the other faces, called *lateral faces*, are parallelograms formed by the straight parallel lines between corresponding vertices of the bases. If the bases have a centre, the line joining the centres is the *axis* of the prism. If the axis is at right angles to the base, the prism is a *right prism* (in which case the lateral faces are rectangles); otherwise it is an *oblique prism*. A *triangular prism* has triangular bases and three lateral faces. This is the shape of many of the glass prisms used in optical instruments. A *quadrangular prism* has a quadrilateral base and four lateral faces. The cube is a special case of this with square bases and square lateral faces.

probability The likelihood of a given event occurring. If an experiment has n possible and equally likely outcomes, m of which are event A, then the probability of A is $P(A) = m/n$. For example, if A is an even number coming up when a die is thrown, then $P(A) = 3/6$. When the probabilities of the different possible results are not already known, and event A has occurred m times in n trials, $P(A)$ is defined as the limit of m/n as n becomes infinitely large.

In set theory, if S is a set of events (called the *sample space*) and A and B are events in S (i.e. subsets of S), the *probability function* P can be represented in set notation. $P(A) = 1$ and $P(0) = 0$ mean that A is 100% certain and the probability of none of the events in S occurring is zero. $0 \leqslant P(A) \leqslant 1$ for all A in S. If A and B are separate *independent* events, i.e., if $A \cap B = 0$, then $P(A \cup B) = P(A) + P(B)$. If $A \cap B \neq 0$ then $P(A \cup B) = P(A) + P(B) - P(A \cap B)$.

The *conditional probability* is the probability that A occurs when it is known that B has occurred. It is written as

$$P(A|B) = P(A \cap B)/P(B)$$

If A and B are independent events, $P(A|B) = P(A)$ and $P(A \cap B) = P(A) P(B)$. If A and B cannot occur simultaneously, i.e. are mutually exclusive events, $P(A \cap B) = 0$.

probability density function *See* random variable.

probability function *See* probability.

procedure *See* subroutine.

processor *See* central processor.

product The result obtained by multiplication of numbers, vectors, matrices, etc. *See also* Cartesian product.

product formulae *See* addition formulae.

program A complete set of instructions to a computer, written in a *programming language.* (The word is also used as a verb, meaning to write such instructions.) These instructions, together with the facts (usually called *data*) on which the instructions operate, enable the computer to perform a wide variety of tasks. For example, there are instructions to do arithmetic, to move data between the main store and the central processor of the computer, to perform logical operations, and to alter the flow of control in the program.

The instructions and data must be expressed in such a way that the central processor can recognize and interpret the instructions and cause them to be carried out on the right data. They must in fact be in binary form, i.e. in a code consisting of the binary digits 0 and 1 (bits). This binary code is known as *machine code* (or *machine language*). Each type of computer has its own machine code.

It is difficult and time-consuming for people to write programs in machine code. Instead programs are usually written in a *source language*, and these *source programs* are then translated into machine code. Most source programs are written in a *high-level language* and are converted into machine code by a complicated program called a *compiler.* High-level languages are closer to natural language and mathematical notation than to machine code, with the instructions taking the form of *statements.* They are fairly easy to use. They are designed to solve particular sorts of problems and are therefore described as 'problem-orientated'. Some of the most common are FORTRAN, ALGOL, BASIC, and PL/1, which are all used for scientific and technical purposes, and COBOL, which is mainly for commercial applications. For each type of computer there will be compilers for a variety of high-level languages.

It is also possible to write source programs in a *low-level language*. These languages resemble machine code more closely than natural language. They are designed for particular computers and are thus described as 'machine-orientated'. *Assembly languages* are low-level languages. A program written in an assembly language is converted into machine code by means of a special program known as an *assembler*. *See also* routine, subroutine, software.

programming language *See* program.

progression *See* sequence.

progressive wave *See* wave.

projectile An object falling freely in a gravitational field, having been projected at a speed v and at an angle of elevation θ to the horizontal. In the special case that $\theta = 90°$, the motion is linear in the vertical direction. It may then be treated using the equations of motion. In all other cases the vertical and horizontal components of velocity must be treated separately. In the absence of friction, the horizontal component is constant and the vertical motion may be treated using the equations of motion. The path of the projectile is an arc of a parabola. Some useful relations are given below.
Time to reach maximum height:
$$t = v\sin\theta/g$$
Maximum height:
$$h = v^2\sin^2\theta/2g$$
Horizontal range:
$$R = v^2\sin2\theta/g$$
See also orbit.

projection A geometrical transformation in which one line, shape, etc., is converted into another according to certain geometrical rules. A set of points (the *object*) is converted into another set (the image) by the projection. *See* central projection, Mercator's projection, orthogonal projection, stereographic projection.

projective geometry The study of how the geometric properties of a figure are altered by projection. There is a one-to-one correspondence between points in a figure and points in its projected image, but often the ratios of lengths will be changed. In central projection for example, a triangle maps into a triangle and a quadrilateral into a quadrilateral, but the sides and angles may change. *See also* projection.

prolate Denoting a spheroid that has a polar diameter that is greater than the equatorial diameter. *Compare* oblate.

proof A logical argument showing that a statement, proposition, or mathematical formula is true. A proof consists of a set of basic assumptions, called axioms or premisses, that are combined according to logical rules, to derive as a conclusion the formula that is being proved. A proof of a proposition or formula P is just a valid argument from true premises to give P as a conclusion. *See also* direct proof, indirect proof.

proper fraction *See* fraction.

proper subset A subset S of a set T is a *proper subset* if there are elements of T that are not in S, i.e. S has fewer elements than T (if T is a finite set).

proportional Symbol: $\propto$ Varying in a constant ratio to another quantity. For example, if the length l of a metal bar increases by 1 millimetre for every 10°C rise in its temperature T, then the length is proportional to temperature and the constant of proportionality k is $1/10$ millimetre per degree Celsius; $l = l_0 + kT$, where l_0 is the initial length. If two quantities a and b are *directly proportional* then $a/b = k$, where k is a constant. If they are *inversely proportional* then their product is a constant; i.e. $ab = k$, or $a = k/b$.

proposition A sentence or formula in a logical argument. A proposition can have a truth value; that is, it can be either true or false but not both. Any logical argument consists of a succession of propositions

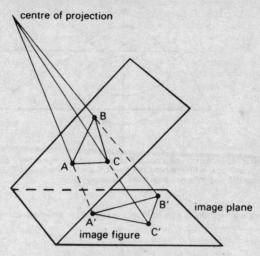

central projection of a triangle ABC
onto a triangle A'B'C'

orthogonal projection of a triangle
ABC onto a triangle A'B'C'

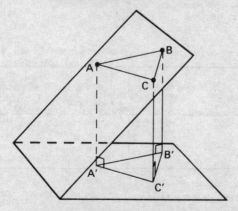

Methods of projecting from
one plane to another

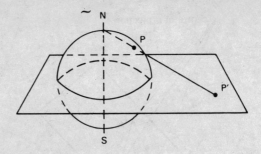

Stereographic projection of a point P on the surface of a sphere onto a plane perpendicular to the line joining the poles N and S. The image P' is obtained by extending the line NP until it crosses the image plane.

Mercator's projection of a point P onto a point P' in an image plane or map.

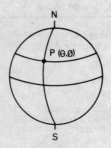

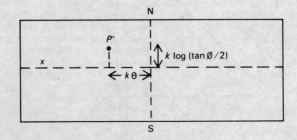

The image P' of a point P in Mercator's projection.

Methods of projecting from the surface of a sphere onto a plane.

148

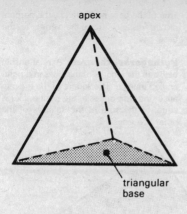

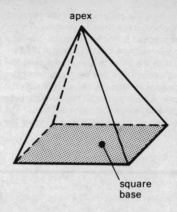

apex

triangular
base

apex

square
base

Triangular pyramid   Square pyramid

linked by logical operations with a proposition as conclusion.

Propositions may be simple or compound. A *compound proposition* is one that is made up of more than one proposition. For example, a proposition P might consist of the constituent parts 'if R, then S or not Q'; i.e. in this case $P = R \rightarrow (S \vee Q)$. A *simple proposition* is one that is not compound. *See also* logic, symbolic logic.

propositional calculus *See* symbolic logic.

propositional logic *See* symbolic logic.

protractor A drawing instrument used for marking out or measuring angles. It usually consists of a semicircular piece of transparent plastic sheet, marked with radial lines at one-degree intervals.

Ptolemy's theorem *See* cyclic polygon.

pulley A class of machine. In any pulley system power is transferred through the tension in a string wound over one or more wheels. The force ratio and distance ratio depend on the relative arrangement

of strings and wheels. The efficiency is not usually very high as work must be done to overcome friction in the strings and the wheel bearings and to lift any moving wheels. *See* machine.

pulsatance *See* angular frequency.

punched card *See* card.

pure mathematics The study of mathematical theory and structures, without necessarily having an immediate application in mind. For example, the study of the general properties of vectors, considered purely as entities with certain properties, could be considered as a branch of pure mathematics. The use of vector algebra in mechanics to solve a problem on forces or relative velocity is a branch of applied mathematics. Pure mathematics, then, deals with abstract entities, without any necessary reference to physical applications in the 'real world'.

pyramid A solid figure in which one of the faces, the base, is a polygon and the others are triangles with the same vertex. If the base has a centre of symmetry, a line from the vertex to the centre is the *axis* of the pyramid. If this axis is at right angles to

Pythagoras' theorem

the base the pyramid is a *right pyramid*; otherwise it is an *oblique pyramid*. A *regular pyramid* is one in which the base is a regular polygon and the axis is at right angles to the base. In a regular pyramid all the lateral faces are congruent isosceles triangles making the same angle with the base. A *square pyramid* has a square base and four congruent triangular faces. The volume of a pyramid is one third of the area of the base multiplied by the perpendicular distance from the vertex to the base.

Pythagoras' theorem A relationship between the lengths of the sides in a right-angled triangle. The square of the hypotenuse (the side opposite the right angle) is equal to the sum of the squares of the other two sides.

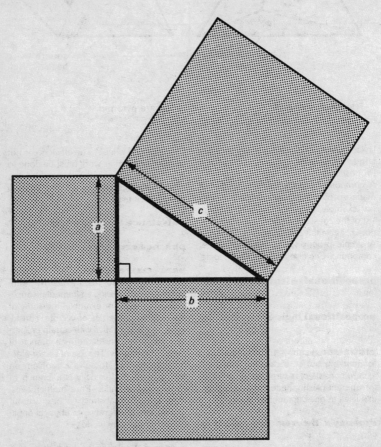

Pythagoras' theorem: $c^2 = a^2 + b^2$

Q

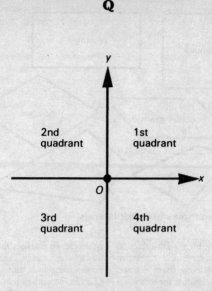

The four quadrants of a two-dimensional Cartesian coordinate system.

quadrangular prism *See* prism.

quadrant 1. One of four divisions of a plane. In rectangular Cartesian coordinates, the first quadrant is the area to the right of the y-axis and above the x-axis, that is, where both x and y are positive. The second quadrant is the area to the left of the y-axis and above the x-axis, where x is negative and y is positive. The third quadrant is below the x-axis and to the left of the y-axis, where both x and y are negative. The fourth quadrant is below the x-axis and to the right of the y-axis, where x is negative and y is positive. In polar coordinates, the first, second, third, and fourth quadrants occur when the direction angle, θ, is 0 to 90° (0 to $\pi/2$); 90° to 180° ($\pi/2$ to π); 180° to 270° (π to $3\pi/2$); and 270° to 360° ($3\pi/2$ to 2π), respectively. *See also* Cartesian coordinates, polar coordinates.

2. A quarter of a circle, bounded by two perpendicular radii and a quarter of the circumference.
3. A unit of plane angle equal to 90 degrees ($\pi/2$ radians). A quadrant is a right angle.

quadrantal spherical triangle *See* spherical triangle.

quadratic equation A polynomial equation in which the highest power of the unknown variable is two. The general form of a quadratic equation in the variable x is

$$ax^2 + bx + c = 0$$

where a, b, and c are constants. It is also sometimes written in the reduced form

$$x^2 + bx/a + c/a = 0$$

In general, there are two values of x that satisfy the equation. These solutions (or roots), are given by the formula

$$x = [-b \pm \sqrt{(b^2 - 4ac)}]/2a$$

151

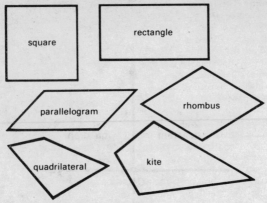

Six examples of quadrilaterals.

The quantity $b^2 - 4ac$ is called the *discriminant*. If it is a positive number, there are two real roots. If it is zero, there are two equal roots. If it is negative there are no real roots. The Cartesian coordinate graph of a quadratic function

$$y = ax^2 + bx + c$$

is a parabola and the points where it crosses the x-axis are solutions to

$$ax^2 + bx + c = 0$$

If it crosses the axis twice there are two real roots, if it touches the axis at a turning point the roots are equal, and if it does not cross it at all there are no real roots. In this last case, where the discriminant is negative, the roots are two conjugate complex numbers. *See also* discriminant.

quadrilateral A plane figure with four straight sides. For example, squares, kites, rhombuses, and trapeziums are all quadrilaterals. A square is a regular quadrilateral.

quantifier *See* existential quantifier, universal quantifier.

quart A unit of capacity equal to two UK pints (in the UK) or two US pints (in the USA). In the USA a dry quart is equal to two US dry pints.

quartic equation A polynomial equation in which the highest power of the unknown variable is four. The general form of a quartic equation in a variable x is

$$ax^4 + bx^3 + cx^2 + dx + e = 0$$

where a, b, c, d, and e are constants. It is also sometimes written in the reduced form

$$x^4 + bx^3/a + cx^2/a + dx/a + e/a = 0$$

In general, there are four values of x that satisfy a quartic equation. For example,

$$2x^4 - 9x^3 + 4x^2 + 21x - 18 = 0$$

can be factorized to

$$(2x + 3)(x - 1)(x - 2)(x - 3) = 0$$

and its solutions (or roots) are $-3/2$, 1, 2, and 3. On a Cartesian coordinate graph, the curve

$$y = 2x^4 - 9x^3 + 4x^2 + 21x - 18 = 0$$

crosses the x-axis at $x = -3/2$; $x = 1$; $x = 2$; and $x = 3$. *Compare* cubic equation, quadratic equation.

quartile One of the three points that divide a set of data arranged in numerical order into four equal parts. The lower (or first) quartile, Q_1, is the 25th percentile (P_{25}). The middle (or second) quartile, Q_2, is the median (P_{50}). The upper (or third) quartile, Q_3, is the 75th percentile (P_{75}). *See also* median, percentile.

quaternions Generalized complex numbers invented by Hamilton. A quaternion is of the form $a + bi + cj + dk$, where $i^2 = j^2 = k^2 = -1$ and $ij = -ji = k$ and a, b, c, d are real numbers. The most striking feature of quaternions is that multiplication is not commutative. They have applications in the study of the rotations of rigid bodies in space.

quintic equation A polynomial equation in which the highest power of the unknown variable is five. The general form of a quintic equation in a variable x is:

$$ax^5 + bx^4 + cx^3 + dx^2 + ex + f = 0$$

where a, b, c, d, e, and f are constants. It is also sometimes written in the reduced form

$$x^5 + bx^4/a + cx^3/a + dx^2/a + ex/a + f/a = 0$$

In general, there are five values of x that satisfy a quintic equation. For example,

$$2x^5 - 17x^4 + 40x^3 + 5x^2 - 102x + 72 = 0$$

can be factorized to

$$(2x + 3)(x - 1)(x - 2)(x - 3)(x - 4) = 0$$

and its solutions (or roots) are $-3/2$, 1, 2, 3, and 4. On a Cartesian coordinate graph, the curve

$$y = 2x^5 - 17x^4 + 40x^3 + 5x^2 - 102x + 72$$

crosses the x-axis at $x = -3/2$; $x = 1$; $x = 2$; $x = 3$; and $x = 4$. *Compare* cubic equation, quadratic equation, quartic equation.

quotient The result of dividing one number by another. There may or may not be a remainder. For example, $16/3$ gives a quotient of 5 and a remainder of 1.

R

radial Along the direction of the radius.

radian Symbol: rad The SI unit for measuring plane angle. It is the angle subtended at the centre of a circle by an arc equal in length to the radius of the circle. π radians $= 180°$.

radical An expression for a root. For example, $\sqrt{2}$, where $\sqrt{}$ is the radical sign.

radical axis The radical axis of two circles
$$x^2 + y^2 + 2a_1x + 2b_1y + c_1 = 0$$
and
$$x^2 + y^2 + 2a_2x + 2b_2y + c_2 = 0$$
is the straight line obtained by eliminating the square terms between the equations of the circles, i.e.
$$2(a_1 - a_2)x + 2(b_1 - b_2)y + (c_1 - c_2) = 0$$
When the circles intersect, the radical axis passes through their two points of intersection.

radius The distance from the centre of a circle to any point on its circumference or from the centre of a sphere to its surface. In polar coordinates, a radius r (distance from a fixed origin) is used with angular position θ to specify the positions of points.

radius of convergence For a power series $a_0 + a_1(x-a) + a_2(x-a)^2 + \ldots + a_n(x-a)^n + \ldots$ there is an R such that the series converges if $|x-a| < R$ and diverges if $|x-a| > R$. R is the radius of convergence of the power series.

radius of curvature See curvature.

radius of gyration Symbol: k For a body of mass m and moment of inertia I about an axis, the radius of gyration about that axis is given by
$$k^2 = I/m$$

In other words, a point mass m rotating at a distance k from the axis would have the same moment of inertia as the body.

radius vector The vector that represents the distance and direction of a point from the origin in a polar coordinate system.

RAM Random-access memory. See random access.

random access A method of organizing information in a computer storage device so that one piece of information may be reached directly in about the same time as any other. Main store, disk units, and drum units all operate by random access and are thus known as *random-access memory* (RAM). In contrast a magnetic tape unit operates more slowly by *serial access*: a particular piece of information can only be retrieved by working through the preceding blocks of data on the tape.

random error See error.

random number table A table consisting of a sequence of randomly chosen digits from 0 to 9, where each digit has a probability of 0.1 of appearing in a particular position, and choices for different positions are independent. Random numbers are used in statistical random sampling.

random sampling See sampling.

random variable (chance variable, stochastic variable) A quantity that can take any one of a number of unpredicted values. A *discrete random variable*, X, has a definite set of possible values $x_1, x_2, x_3, \ldots x_n$, with corresponding probabilities $p_1, p_2, p_3, \ldots p_n$. Since X must take one of the values in this set,
$$p_1 + p_2 + \ldots + p_n = 1$$
If X is a *continuous random variable*, it can take any value over a continuous range. The probabilities of a particular value x oc-

curring is given by a *probability density function* f(x). On a graph of f(x) against x, the area under the curve between two values a and b is the probability that X lies between a and b. The total area under the curve is 1.

random walk A succession of movements along line segments where the direction and the length of each move is randomly determined. The problem is to determine the probable location of a point subject to such random motions given the probability of moving some distance in some direction, where the probabilities are the same at each step. Random walks can be used to obtain probability distributions to practical problems. Consider, for example, a drunk man moving a distance of one unit in unit time, the direction of motion being random at each step. The problem is to find the probability distribution of the distance of the point from the starting point after some fixed time. Technically a random walk is a sequence

$$S_n = X_1 + X_2 + \ldots + X_n$$

where $\{X_i\}$ is a sequence of independent random variables.

range 1. The difference between the largest and smallest values in a set of data. It is a measure of dispersion. In terms of percentiles, the range is $(P_{100} - P_0)$. *Compare* interquartile range.
2. A set of numbers or quantities that form possible results of a mapping. In algebra, the range of a function f(x) is the set of values that f(x) can take for all possible values of x. For example, if f(x) is taking the square root of positive rational numbers, then the range would be the set of real numbers. *See also* domain.

rank A method of ordering a set of objects according to the magnitude or importance of a variable measured on them, e.g. arranging ten men in order of height. If the objects are ranked using two different variables, the degree of association between the two rankings is given by the coefficient of rank correlation. *See also* Kendall's method.

ratio One number or quantity divided by another. The ratio of two variable quantities x and y, written as x/y or $x:y$, is constant if y is proportional to x. *See also* fraction.

rationalized units A system of units in which the equations have a logical form related to the shape of the system. SI units form a rationalized system of units. For example, in it formulae concerned with circular symmetry contain a factor of 2π; those concerned with radial symmetry contain a factor of 4π.

rational numbers Symbol: Q The set of numbers that includes integers and fractions. Rational numbers can be written down exactly as ratios or as finite or repeating decimals. For example, $1/3$ ($= 0.333\ldots$) and $1/4$ ($= 0.25$) are rational. The square root of $2 (= 1.414\,213\,6\ldots)$ is not. *Compare* irrational numbers.

ray A set consisting of all the points on a given line to the left or to the right of a given point, and including that point itself.

reaction Newton's third law of force states that whenever object A applies a force on object B, B applies the same force on A. An old word for force is 'action'; 'reaction' is thus the other member of the pair. Thus in the interaction between two electric charges, each exerts a force on the other. Thus, in general, action and reaction have little meaning. The word 'reaction' is still sometimes used in restricted cases, such as the reaction of a support on the object it supports. In this case the 'action' is the effect of the weight of the object on the support.

reader A device used in a computer system to sense the information recorded on some source and convert it into another form. A paper-tape reader, for example, senses the series of holes punched in a paper tape and converts the information into a series of electrical pulses, which can be transmitted to the central processor of the computer. *See also* card, OCR.

read-only memory (ROM) *See* store.

read–write head *See* disk, drum, magnetic tape.

real numbers Symbol: R The set of numbers that includes all rational and irrational numbers.

real time The actual time in which a physical process takes place or in which a physical process, machine, etc., is under the direct control of a computer. A *real-time system* is able to react sufficiently rapidly so that it may control a continuing process, making changes or modifications when necessary. Air-traffic control and airline reservations require real-time systems. *Compare* batch processing. *See also* time sharing.

reciprocal The number 1 divided by a quantity. For example, the reciprocal of 2 is $\frac{1}{2}$. The reciprocal of $(x^2 + 1)$ is $1/(x^2 + 1)$. The product of any expression and its reciprocal is 1. For any function, the reciprocal is the multiplicative inverse.

rectangle A plane figure with four straight sides, two parallel pairs of equal length forming four right angles. The area of a rectangle is the product of the two different side lengths, the length times the breadth. A rectangle has two axes of symmetry, the two lines joining the mid-points of opposite sides. It can also be superimposed on itself after rotation through $180°$ (π radians). The two diagonals of a rectangle have equal lengths.

rectangular hyperbola *See* hyperbola.

rectangular parallelopiped *See* parallelopiped.

rectilinear Describing motion in a straight line.

recurring decimal A repeating decimal. *See* decimal.

reduced form (of a polynomial) The equation of the form
$$x^n + (b/a)x^{n-1} + (c/a)x^{n-2} + \ldots = 0$$
that is derived from a polynomial of the form
$$ax^n + bx^{n-1} + cx^{n-2} + \ldots = 0$$
For example,
$$2x^2 - 10x + 12 = 0$$
is equivalent to the reduced form
$$x^2 - 5x + 6 = 0$$
See also equation, polynomial, quadratic equation.

reductio ad absurdum A method of proof which proceeds by assuming that falsity of what we wish to prove and showing that it leads to a contradiction. Hence the statement whose falsity we assumed must be true. The following proof that $\sqrt{2}$ is irrational is a simple example of proof by this method.
Assume $\sqrt{2}$ is rational. In that case it can be expressed in the form a/b where a and b are integers. Assume that this fraction is in its lowest terms and so a and b have no common factor. Since $a/b = \sqrt{2}$ then $a^2/b^2 = 2$. Hence $a^2 = 2b^2$. This means that a^2 is even and hence a itself is even. In that case we can write a as $2m$ where m is some integer. But then since $a^2 = 2b^2$ we have $(2m)^2 = 2b^2$, or $4m^2 = 2b^2$. Dividing by 2 we get $2m^2 = b^2$. But this means that b^2 and hence b is also even. Hence, a and b do have a common factor, namely 2. But we assumed they had no common factor. Since we have reached a contradiction, our starting-point – the assumption that $\sqrt{2}$ is rational – must be false.

reduction formulae In trigonometry, the equations that express sine, cosine, and tangent functions of an angle in terms of an angle between 0 and $90°$ ($\pi/2$). For example:
$$\sin(90° + \alpha) = \cos\alpha$$
$$\sin(180° + \alpha) = -\sin\alpha$$
$$\sin(270° + \alpha) = -\cos\alpha$$
$$\cos(90° + \alpha) = -\sin\alpha$$
$$\tan(90° + \alpha) = -\cot\alpha$$

reflection The geometrical transformation of a point or a set of points from one side of a point, line, or plane to a symmetrical position on the other side. On *reflection in a line*, the image of a point P would be point P' at the same distance from the line but on the other side. The line, the *axis of reflection*, is the perpendicular bisector of the line PP'. In a symmetrical plane figure there is an axis of reflection, also called the *axis of symmetry*, in which the figure is reflected onto itself. An equilateral triangle, for example, has three axes of symmetry. In a circle, any diameter is an axis of symmetry. Similarly, a solid may undergo *reflection in a plane*. In a sphere, any plane passing through the sphere's centre would be a plane of symmetry.

In a Cartesian coordinate system, reflection in the x-axis changes the sign of the y-coordinate. A point (a,b) would become $(a,-b)$. Reflection in the y-axis changes the sign of the x-coordinate, making (a,b) become $(-a,b)$. In three dimensions, changing the sign of the z-coordinate is equivalent to reflection in the plane of the x and y axes. *Reflection in a point* is equivalent to rotation through $180°$. Each point P is moved to a position P' so that the point of reflection bisects the line PP'. *Reflection in the origin* of plane Cartesian coordinates changes the signs of all the coordinates. It is equivalent to reflection in the x-axis followed by reflection in the y-axis, or vice versa. *See also* rotation.

reflexive A relation R defined on a given set is said to be reflexive if every member of the set has this relation to itself. Equality is an example of a reflexive relation.

register *See* central processor.

regression line A line $y = ax + b$, called the regression line of y on x, which gives the expected value of a random variable y conditional on the given value of a random variable x. The regression line of x on y is not in general the same as that of y on x. If a scatter diagram of data points (x_1,y_1), $\ldots$, (x_n,y_n) is drawn and a linear relationship is shown up, the line can be drawn by hand. The best line is drawn using the least-squares method. *See also* correlation, least squares method, scatter diagram.

regular Having all faces or sides of equal size and shape. *See* polygon, polyhedron.

relation A property that holds for ordered pairs of elements of some set, for example being greater than. We can think of a relation abstractly as the set of all ordered pairs in which the two members have the given relation to one another.

relative Expressed as a difference from or as a ratio to, some reference level. Relative density, for example, is the mass of a substance per unit volume expressed as a fraction of a standard density, such as that of water at the same temperature. *Compare* absolute.

relative error The error or uncertainty in a measurement expressed as a fraction of the measurement. For example, if, in measuring a length of 10 metres, the tape measures only to the nearest centimetre, then the measurement might be written as 10 ± 0.01 metres. The relative error is $0.01/10 = 0.001$. *Compare* absolute error. *See also* error.

relative maximum *See* local maximum.

relative minimum *See* local minimum.

relative velocity If two objects are moving at velocities v_A and v_B in a given direction the velocity of A relative to B is $v_A - v_B$ in that direction. In general, if two objects are moving in the same frame at nonrelativistic speeds their relative velocity is the vector difference of the two velocities.

relativistic mass The mass of an object as measured by an observer at rest in a frame of reference in which the object is moving with a velocity v. It is given by
$$m_0 = m\sqrt{(1 - v^2/c^2)}$$
where m_0 is the rest mass, c is the velocity of light, and m is the relativistic mass. The equation is a consequence of the special theory of relativity, and is in excellent agreement with experiment. No object

can travel at the speed of light because its mass would then be infinite. *See also* relativity, theory of; rest mass.

relativistic mechanics A system of mechanics based on relativity theory. *See also* classical mechanics.

relativistic speed (relativistic velocity) Any speed (velocity) that is sufficiently high to make the mass of an object significantly greater than its rest mass. It is usually expressed as a fraction of c, the speed of light in free space. At a speed of $c/2$ the relativistic mass of an object is about 15% greater than the rest mass. *See also* relativistic mass, rest mass.

relativity, theory of A theory put forward in two parts by Albert Einstein. The special theory (1905) referred only to nonaccelerated (inertial) frames of reference. The general theory (1915) is also applicable to accelerated systems.

The *special theory* was based on two postulates:

(1) That physical laws are the same in all inertial frames of reference.

(2) That the speed of light in a vacuum is constant for all observers, regardless of the motion of the source or observer.

The second postulate seems contrary to 'common sense' ideas of motion. Einstein was led to the theory by considering the problem of the 'ether' and the relation between electric and magnetic fields in relative motion. The theory accounts for the negative result of the Michelson–Morley experiment and shows that the Lorentz–Fitzgerald contraction is only an apparent effect of motion on an object relative to an observer, not a 'real' contraction. It leads to the result that the mass of an object moving at a speed v relative to an observer is given by:

$$m = m_0/\sqrt{(1 - v^2/c^2)}$$

where c is the speed of light and m_0 the mass of the object when at rest relative to the observer. The increase in mass is significant at high speeds. Another consequence of the theory is that an object has an energy content by virtue of its mass, and similarly that energy has inertia. Mass and energy are related by the famous equation $E = mc^2$.

The *general theory* of relativity seeks to explain the difference between accelerated and nonaccelerated systems and the nature of the forces acting in both of them. For example, a person in a spacecraft far out in space would not be subject to gravitational forces. If the craft were rotating, he would be pressed against the walls of the craft and would consider that he had weight. There would not be any difference between this force and the force of gravity. To an outside observer the force is simply a result of the tendency to continue in a straight line; i.e. his inertia. This type of analysis of forces led Einstein to a *principle of equivalence* that inertial forces and gravitational forces are equivalent, and that gravitation can be explained by the geometrical properties of space. He visualized a four-dimensional space–time continuum in which the presence of a mass affects the geometry – the space is 'curved' by the mass.

remainder The number left when one number is divided into another. Dividing 12 into 57 gives 4 remainder 9 ($4 \times 12 = 48$; $57 - 48 = 9$).

remainder theorem The theorem expressed by the equation

$$f(x) = (x - a)g(x) + f(a)$$

This means that if a polynomial in x, f(x), is divided by $(x - a)$, where a is a constant, the remainder term is equal to the value of the polynomial when $x = a$. For example, if

$$2x^3 + 3x^2 - x - 4$$

is divided by $(x - 4)$, then the remainder term is

$$f(4) = 128 + 48 - 4 - 4 = 168$$

The remainder theorem is useful for finding the factors of a polynomial. In this example,

$$f(1) = 2 + 3 - 1 - 4 = 0$$

Thus, there is no remainder so $(x - 1)$ is a factor.

remote job entry *See* batch processing.

repeating decimal *See* decimal.

representative fraction A fraction used to express the scale of a map in which the numerator represents a distance on the map and the denominator represents the corresponding distance on the ground. As a fraction is a ratio, the units of the numerator and denominator must be the same. For example, a scale of 1 cm = 1 km would be given as a representative fraction of 1/100 000, because there are 100 000 cm in 1 km. *See also* scale.

residue If there exists an x such that $x^n \equiv a \pmod{p}$, i.e. x^n is congruent to a modulo p, then a is called a residue of p of order n.

resolution of vectors The determination of the components of a vector in two given directions at 90°. The term is sometimes used in relation to finding any pair of components (not necessarily at 90° to each other).

resonance The large-amplitude vibration of an object or system when given impulses at its natural frequency. For instance, a pendulum swings with a natural frequency that depends on its length and mass. If it is given a periodic 'push' at this frequency – for example, at each maximum of a complete oscillation – the amplitude is increased with little effort. Much more effort would be required to produce a swing of the same amplitude at a different frequency.

restitution, coefficient of Symbol: e For the impact of two bodies, the elasticity of the collision is measured by the coefficient of restitution. It is the relative velocity after collision divided by the relative velocity before collision (velocities measured along the line of centres). For spheres A and B:

$$v_A' - v_B' = e(v_A - v_B)$$

v indicates velocity before collision; v' velocity after collision. Kinetic energy is conserved only in a perfectly elastic collision.

rest mass Symbol: m_0 The mass of an object at rest as measured by an observer at rest in the same frame of reference. *See also* relativistic mass.

resultant 1. A vector with the same effect as a number of vectors. Thus, the resultant of a set of forces is a force that has the same effect; it is equal in magnitude and opposite in direction to the equilibrium. Depending on the circumstances, the resultant of a set of vectors can be found by different methods. *See* parallel forces, parallelogram of vectors, principle of moments.
2. *See* eliminant.

revolution, solid of A solid generated by revolving a plane area about a line called the *axis of revolution*. For example, rotating a rectangle about an axis joining the midpoints of two opposite sides produces a cylinder as the solid of revolution.

rhombohedron A solid figure bounded by six faces, each one a parallelogram, with opposite faces congruent.

rhomboid A parallelogram that is neither a rhombus nor a rectangle. *See* parallelogram.

rhombus A plane figure with four straight sides of equal length; i.e. a parallelogram with equal sides. Its area is equal to half the product of the lengths of its two diagonals, which bisect each other perpendicularly. The rhombus is symmetrical about both of its diagonals and also has rotational symmetry, in that it can be superimposed on itself after rotation through 180° (π radians).

Riemann integral *See* definite integral, Riemann sum.

Riemann sum The series that approximates the area between the curve of a function f(x) and the x-axis:

$$\sum_{i=1}^{n} f(\xi_i)\Delta x_i$$

where Δx is an increment of x, ξ_i is any value of $f(x)$ within that interval, and n is the number of intervals. The definite (or Riemann) integral is the limit of the sum as n becomes infinitely large and Δx infinitesimally small.

right angle An angle that is $90°$ or $\pi/2$ radians. It is the angle between two lines or planes that are perpendicular to each other. The corner of a square, for example, is a right angle.

right solid A solid geometrical figure that is upright; for example, a cone, cylinder, pyramid, or prism that has an axis at right angles to the base. *Compare* oblique solid.

rigid body In mechanics, a body for which any change of shape produced by forces on the body can be neglected in the calculations.

ring A set of entities with two binary operations called addition and multiplication and denoted by $+$ and $\cdot$ respectively, such that:
(1) the set is a commutative group under addition;
(2) for every pair of elements a,b, the product $a \cdot b$ is unique, multiplication is associative, i.e. $(a \cdot b) \cdot c = a \cdot (b \cdot c)$, and multiplication is distributive with respect to addition, i.e. $a \cdot (b + c) = a \cdot b + a \cdot c$ and $(b + c) \cdot a = b \cdot a + c \cdot a$ for each a, b, and c in the set.
If multiplication is also commutative, the ring is called a *commutative ring*. For example, the set of real numbers, the set of integers, and the set of rational numbers are rings with respect to ordinary addition and multiplication.

robotics A feedback-controlled mechanical device. *Robotics* is the study of the design, applications, and control and sensory systems of robots; for example, the design of robot arms that can approach an object from any orientation and grip it. A robot's control system may be simple and consist of only a sequencing device so that the device moves in a repetitive pattern, or

more sophisticated so that the robot's movements are generated by computer from data about the environment. The robot's sensory system gathers information needed by the control system, usually visually by using a television camera.

Rolle's theorem A curve that intersects the x-axis at two points a and b, is continuous, and has a tangent at every point between a and b, must have at least one point in this interval at which the tangent to the curve is horizontal. For a curve $y = f(x)$, it follows from Rolle's theorem that the function $f(x)$ has a turning point (a maximum or minimum value) between $f(a)$ and $f(b)$, where the derivative $f'(x) = 0$. *See also* turning point.

rolling friction *See* friction.

Roman numerals The system of writing integers used by the Romans in which I denotes 1, V denotes 5, C denotes 100, D denotes 500, and M denotes 1000. The integers are written using the following rules:
(1) the values of the letters are added if a letter is repeated or immediately followed by a letter of lesser value;
(2) the value of the letter of smallest value is subtracted from the value of the letter of greater value when a letter is immediately followed by a letter of greater value.
There is no symbol for zero. The integers from 1 to 10 are written I, II, III, IV, V, VI, VII, VIII, IX, X; and, for example, 1987 is written MCMLXXXVII.

root In an equation, a value of the independent variable that satisfies the equation. In general, the degree of a polynomial is equal to the number of roots. A quadratic equation (one of degree two) has two roots, although in some circumstances they may be equal. For a number a, an nth root of a is a number that satisfies the equation

$$x^n = a$$

See also discriminant, polynomial, quadratic equation.

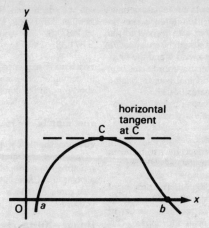

Rolle's theorem for a function f(x) that is continuous between x = a and x = b and for which f(a) = f(b) = 0.

rose A curve obtained by plotting the equation

$$r = a\sin n\theta$$

in polar coordinates (a is a real-number constant and n is an integer constant). It has a number of petal-shaped loops, or leafs. When n is even there are 2n loops and when n is odd there are n loops. For example, the graph of $r = a\sin 2\theta$ is a four-leafed rose.

rotation A geometrical transformation in which a figure is moved rigidly around a fixed point. If the point, the centre of rotation, is labelled O, then for any point P in the figure, moving to point P' after rotation, the angle POP' is the same for all points in the figure. This angle is the angle of rotation. Some figures are unchanged by certain rotations. A circle is not affected by any rotation about its centre. A square does not change if it is rotated through 90° about the point at which its diagonals cross. Similarly an equilateral triangle is unchanged by rotation through 120° about its centroid. These properties are known as the *rotational symmetry* of the

figure. *See also* rotation of axes, transformation.

rotational motion Motion of a body turning about an axis. The physical quantities and laws used to describe linear motion all have rotational analogues; the equations of rotational motion are the analogues of the equations of motion (linear).

As well as the kinematic equations, the equations of rotational motion include $T = I\alpha$, the analogue of $F = ma$. Here T is the turning-force, or torque (the analogue of force), I is the moment of inertia (analogous to mass), and α is the angular acceleration (analogous to linear acceleration). The kinematic equations relate the angular velocity, ω_1, of the object at the start of timing to its angular velocity, ω_2, at some later time, t, and thus to the angular displacement ϕ. They are:

$$\omega = \omega_1 + \alpha t$$
$$\theta = (\omega_1 + \omega_2)/2t$$
$$\theta = \omega_1 t + \alpha t^2/2$$
$$\theta = \omega_2 t - \alpha t^2/2$$
$$\omega_2{}^2 = \omega_1{}^2 + 2\alpha\theta$$

rotation of axes In coordinate geometry, the shifting of the reference axes so that they are rotated with respect to the original axes of the system by an angle (θ). If the new axes are x' and y' and the original axes x and y, then the coordinates (x,y) of a point with the original axes are related to the new coordinates (x',y') by:
$$x = x'\cos\theta - y'\sin\theta$$
$$y = x'\sin\theta - y'\cos\theta$$

rough In mechanics, describing a system in which frictional effects have to be taken into consideration in the calculations.

rounding (rounding off) The process of adjusting the least significant digit or digits in a number after a required number of digits has been truncated (dropped). This reduces the error arising from truncation but still leaves a *rounding error* so that the accuracy of, say, the result of a calculation

will be decreased. For example, the number 2.871 329 71 could be truncated to 2.871 32 but would be rounded to 2.871 33.

routine A sequence of instructions used in computer programming. It may be a short program or sometimes part of a program. *See also* subroutine.

row matrix *See* row vector.

row vector (row matrix) A number, (n), of quantities arranged in a row; i.e. a $1 \times n$ matrix. For example, the coordinates of a point in a Cartesian coordinate system with three axes is a 1×3 row vector, (x,y,z).

Runge–Kutta method An iterative technique for solving ordinary differential equations, used in computer analysis. *See also* differential equation, iteration.

S

saddle point A stationary point on a curved surface, representing a function of two variables, f(x,y), that is not a turning point, i.e. it is neither a maximum nor a minimum value of the function. At a saddle point, the partial derivatives $\partial f/\partial x$ and $\partial f/\partial y$ are both zero, but do not change sign. The tangent plane to the surface at the saddle point is horizontal. Around the saddle point the surface is partly above and partly below this tangent plane.

sample space *See* probability.

sampling The selection of a representative subset from a whole population. Analysis of the sample gives information about the whole population. This is called *statistical inference*. For example, population parameters (such as the population mean and variance) can be estimated using sample statistics (such as the sample mean and variance). Significance (or hypothesis) tests are used to test whether observed differences between two samples are due to chance variation or are significant, as in testing a new production process against an old. The population can be finite or infinite. In sampling with replacement, each individual chosen is returned to the population before the next choice is made. In *random sampling* every member of the population has an equal chance of being chosen. In *stratified random sampling* the population is divided into strata and the random samples drawn from each are pooled. In *systematic sampling* the population is ordered, the first individual chosen at random and further individuals chosen at specified intervals, for example, every 100th person on the electoral roll. If a random sample of size n is the set of numerical values $\{x_1, x_2, \ldots x_n\}$, the sample mean is:

$$\sum_1^n \overline{x} = x_i/n$$

The sample variance is:
$$\Sigma(x_i - \overline{x})^2/(n - 1)$$

$$\Sigma(x_i - \overline{x})^2/n$$
for a normal distribution. If μ is the population mean, the sample variance is:
$$\Sigma(x_i - \mu)/n$$
See also hypothesis test.

sampling distribution The distribution of a sample statistic. For example, when different samples of size n are taken from the same population the means of each sample form a sampling distribution. If the population is infinite (or very large) and sampling is with replacement, the mean of the sample means is $\mu_x = \mu$ and the standard deviation of the sample means is $\sigma_x = \sigma/\sqrt{n}$, where μ and σ are the population mean and standard deviation. When $n \geqslant 30$, sampling distributions are approximately normal and large-sampling theory is used. When $n < 30$, exact sample theory is used. *See also* sampling.

satisfy To be a solution of. For example, 3 and −3 satisfy the equation $x^2 = 9$.

scalar A number or a measure in which direction is unimportant or meaningless. For instance, distance is a scalar quantity, whereas displacement is a vector. Mass, temperature, and time are scalars – they are each quoted as a pure number with a unit. *See also* vector.

scalar product A multiplication of two vectors to give a scalar. The scalar produt of **A** and **B** is defined by $\mathbf{A}.\mathbf{B} = AB\cos\theta$, where A and B are the magnitudes of **A** and **B** and θ is the angle between the vectors. An example is a force **F** displaced **s**. Here the scalar product is energy transferred (or work done):
$$W = \mathbf{F}.\mathbf{s}$$
$$W = Fs\cos\theta$$
where θ is the angle between the line of action of the force and the displacement. A scalar product is indicated by a dot between the vectors and is sometimes called a *dot product*. The scalar product is commutative

$$A.B = B.A$$

and is distributive with respect to vector addition

$$A.(B + C) = A.B + A.C$$

If A is perpendicular to B, $A.B = 0$. In two-dimensional . Cartesian coordinates with unit vectors i and j in the x- and y- directions respectively,

$$A.B = (a_1i + a_2j).(b_1i + b_2j) =$$
$$a_1b_1 + a_2b_2$$

See also vector product.

scalar projection The length of an orthogonal projection of one vector on another. For example, the scalar projection of A on B is $A\cos\theta$, or $(A.B)/b$ where θ is the smaller angle between A and B and b is the unit vector in the direction of B. Compare vector projection.

scale 1. The markings on the axes of a graph, or on a measuring instrument, that correspond to values of a quantity. Each unit of length on a linear scale represents the same interval. For example, a thermometer that has markings 1 millimetre apart to represent 1°C temperature intervals has a linear scale. See also logarithmic scale.
2. The ratio of the length of a line between two points on a map to the distance represented. For example, a map in which two points 5 kilometres apart are shown 5 centimetres apart has a scale of 1/100 000.

scale factor The multiplying factor for each linear measurement of an object when it is to be enlarged about a given centre of enlargement. A scale factor can be positive or negative. If the scale factor is positive, the image is on the same side of the centre of enlargement as the object. If the scale factor is negative, the image will be on the opposite side of the centre of enlargement and will be inverted. Fractional scale factors can be used – these give images that are smaller than the objects.

scalene Denoting a triangle with three unequal sides.

scatter diagram (Galton graph) A graphical representation of data from a bivariate distribution (x,y). The variables are measured on n individuals giving data (x_2,y_1), ..., (x_n,y_n); e.g. x_i and y_i are the height and weight of the ith individual. If y_i is plotted against x_i the resultant scatter diagram will indicate any relationship between x and y by showing if a smooth curve can be drawn through the points. If the points seem to lie near a straight line they are said to be linearly correlated. If they lie near another type of curve they are non-linearly correlated. Otherwise they are uncorrelated. See also regression line.

scientific notation (standard form) A number written as the product of a number between 1 and 10 with a power of 10. For example, 2342.6 in scientific notation is $2.342\ 6 \times 10^3$, and 0.0042 is written as 4.2×10^{-3}.

screw A type of machine, related to the inclined plane, and, in practice, to the second-order lever. The efficiency of screw systems is very low because of friction. Even so, the force ratio (F_2/F_1) can be very high. The distance ratio is given by $2\pi r/p$, where r is the radius and p the pitch of the screw (the angle between the thread and a plane at right angles to the barrel of the screw).

scruple A unit of mass equal to 20 grains. It is equivalent to 1.295 978 grams.

sec See secant.

secant 1. A line that intersects a curve. The intercept is a chord of the curve.
2. (sec) A trigonometric function of an angle equal to the reciprocal of its cosine; i.e. $\sec\alpha = 1/\cos\alpha$. See also trigonometry.

sech A hyperbolic secant. See hyperbolic functions.

second 1. Symbol: s The SI unit of time. It is defined as the duration of 9 192 631 770 cycles of a particular wavelength of radiation corresponding to a transition between two hyperfine levels in the ground state of the caesium-133 atom. **2.** A unit of plane angle equal to one three-hundred and sixtieth of a degree.

second-order determinant *See* determinant.

second-order differential equation A differential equation in which the highest derivative of the dependent variable is a second derivative. *See* differential equation.

section *See* cross section.

sector Part of a circle formed between two radii and the circumference. The area of a sector is $^1/_2 r^2\theta$, where r is the radius and θ is the angle, in radians, formed at the centre of the circle between the two radii.

segment Part of a line or curve between two points, part of a plane figure cut off by a straight line, or part of a solid cut off by a plane. For example, on a graph, a line segment may show the values of a function within a certain range. The area between a chord of a circle and the corresponding arc is a segment of the circle. A cut through a cube parallel to one of its faces forms two cuboid segments.

semantics In computing or logic, the semantics of a language or system is the meaning of particular forms or expressions in the language or system.

semicircle Half a circle, bounded by a diameter and half of the circumference.

semiconductor memory *See* store.

semigroup A set G with a binary operation, $\cdot$, that maps $G \times G$, the set of ordered pairs of members of G, into G, and that is associative, i.e. $a\cdot(b\cdot c) = (a\cdot b)\cdot c$ for all a, b, and c in G. A semigroup is *Abelian* or *commutative* if $a\cdot b = b\cdot a$ for all a,b in G. Sometimes a cancellation law is assumed, i.e. $x = y$ if there is an element z such that $xz = yz$ or $zx = zy$. *Compare* group.

semilogarithmic graph *See* log-linear graph.

separation of variables A method of solving ordinary differential equations. In a first order differential equation,
$$dy/dx = F(x,y)$$
if $F(x,y)$ can be written as $f(x),g(y)$, the variables in the function are separable and the equation can therefore be solved by writing it as
$$dy/g(y) = f(x)dx$$
and integrating both sides. For example
$$dy/dx = x^2y$$
can be written
$$(1/y)dy = x^2dx$$
See also differential equation.

sequence (progression) An ordered set of numbers. Each term in a sequence can be written as an algebraic function of its position. For example, in the sequence (2, 4, 6, 8, ...) the general expression for the nth term is $a_n = 2^n$. A *finite sequence* has a definite number of terms. An *infinite sequence* has an infinite number of terms. *Compare* series. *See also* arithmetic sequence, geometric sequence, convergent sequence, divergent sequence.

serial access *See* random access.

series The sum of an ordered set of numbers. Each term in the series can be written as an algebraic function of its position. For example, in the series $2 + 4 + 6 + 8$ ... the general expression for the nth term, a_n, is 2^n. A *finite series* has a finite fixed number of terms. An *infinite series* has an infinite number of terms. A series with m terms, or the sum of the first m terms of an infinite series, can be written as S_m or
$$\Sigma a_n$$
Compare sequence. *See also* arithmetic series, geometric series, convergent series, divergent series.

set Any collection of things or numbers that belong to a well-defined category. For example, 'dog' is a member, or *element*, of the set of 'types of four-legged animal'. The set of 'days in the week' has seven elements. In a set notation, this would be written as {Monday, Tuesday, ... } = 7. This kind of set is a *finite set*. Some sets, such as the set of natural numbers, N = {1, 2, 3, ... }, have an infinite number of elements. A line segment also is an *infinite set* of points.

Another way of writing a set of numbers is by defining it algebraically. The set of all numbers between 0 and 10 could be written as {x:0$<x<$10}. That is, all values of a variable, x, such that x is greater than zero and less than ten. *See also* Venn diagram.

set square A drawing instrument, consisting of a flat right-angled triangular shape, used for drawing right angles and angles of 30°, 45°, and 60°. Some types have a moving part with a scale so that other angles can be drawn.

sexagesimal Based on multiples of 60. The measurement of an angle in degrees, minutes, and seconds, for example, is a sexagesimal measure, because there are 60 seconds in one minute, and 60 minutes in one degree. A sexagesimal number is one that uses 60 as a base instead of 10. *See also* base.

sextant A unit of plane angle equal to 60 degrees ($\pi/3$ radians).

sextic An equation of degree six, i.e.
$$ax^6 + bx^5 + cx^4 + dx^3 + ex^2 + fx + g = 0$$
There is no general method of solution for such an equation.

shares *See* stocks and shares.

sheaf A sheaf of planes is a set of planes that all pass through a given point, called the *centre* of the sheaf. *Compare* pencil.

s.h.m. *See* simple harmonic motion.

SI *See* SI units.

siemens (mho) Symbol: S The SI unit of electrical conductance, equal to a conductance of one ohm^{-1}.

sieve of Eratosthenese *See* Eratosthenese, sieve of.

sign A unit of plane angle equal to 30 degrees ($\pi/6$ radian).

signed number A number that is denoted as positive or negative.

significance test *See* hypothesis test.

significant figures The number of digits used to denote an exact value to a specified degree of accuracy. For example, 6084.324 is a value accurate to seven significant figures. If it is written as approximately 6080, it is accurate to three significant figures The final 0 is not significant because it is used only to show the order of magnitude of the number.

similar Denoting two or more figures that differ in scale but not in shape. The conditions for two triangles to be similar are:
(1) Three sides of one are proportional to three sides of the other.
(2) The angle of one is equal to the angle of the other and the sides forming the angle in one proportional to the same sides in the other.
(3) Three angles of one are equal to three angles of the other.
Compare congruent.

simple harmonic motion (s.h.m.) Any motion that can be drawn as a sine wave. Examples are the simple oscillation (vibration) of a pendulum or a sound source and the variation involved in a simple wave motion. Simple harmonic motion is observed when the system, moved away from the central position, experiences a restoring force that is proportional to the displacement from this position.

The equation of motion for such a system can be written in the form:
$$md^2x/dt^2 = \lambda x$$
λ being a constant. During the motion there is an exchange of kinetic and potential energy, the sum of the two being constant (in the absence of damping). The period (T) is given by

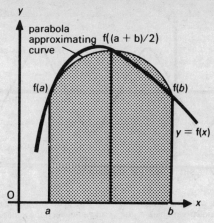

The Simpson's rule approximation for the area under a curve $y = f(x)$, using two columns in the interval $x = a$ to $x = b$.

$$T = 1/f$$

$$T = 2\pi/\omega$$

where f is frequency and ω pulsatance. Other relationships are:

$$x = x_0 \sin\omega t$$
$$dx/dt = \pm\omega\sqrt{(x_0{}^2 - x^2)}$$
$$d^2x/dt^2 = -\omega^2 x$$

Here x_0 is the maximum displacement; i.e. the amplitude of the vibration. In the case of angular motion, as for a pendulum, θ is used rather than x.

A simple harmonic motion can be represented by the motion of a point at constant speed in a circular path. The foot of the perpendicular from the point to an axis through a diameter describes a simple harmonic motion. This is used in a method of representing simple harmonic motions by rotating vectors (called *phasors*).

simple interest The interest earned on capital when the interest is withdrawn as it is paid, so that the capital remains fixed. If the amount of money invested (the principal) is denoted by P, the time in years by T, and the percentage rate per annum by R, then the simple interest is $PRT/100$. *Compare* compound interest.

simple proposition *See* proposition.

Simpson's rule A rule for finding the approximate area under a curve by dividing it into pairs of vertical columns of equal width with bases lying along the horizontal axis. Each pair of columns is bounded by the vertical lines from the x-axis to three corresponding points on the curve and at the top by a parabola that goes through these three points, which approximates the curve. For example, if the value of $f(x)$ is known at $x = a$, $x = b$, and at a value mid-way between a and b the integral of $f(x)dx$ between limits a and b is approximately equal to:

$$h\{f(a) + 4f[(a+b)/2] + f(b)\}/3$$

h is half the distance between a and b. As with the trapezium rule, which is less accurate, better approximations can be obtained by subdividing the area into 4, 6, 8, ... columns until further subdivision makes no significant difference to the result. *Compare* trapezium rule. *See also* numerical integration.

simultaneous equations A set of two or more equations that together speci-

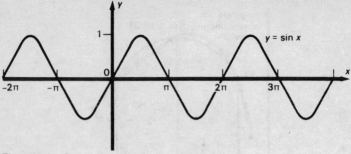

The graph of $y = \sin x$, with x in radians.

fy conditions for two or more variables. If the number of unknown variables is the same as the number of equations, then there is a unique value for each variable that satisfies all the equations. For example, the equations

$$x + 2y = 6$$

and

$$3x + 4y = 9$$

have the solution $x = -3; y = -4.5$. The method of solving simultaneous equations is to eliminate one of the variables by adding or subtracting the equations. For example, multiplying the first equation above by 2 and subtracting it from the second gives:

$$3x + 4y - 2x - 4y = 9 - 12$$

i.e. $x = -3$. Substituting this into either equation gives the value of y. Simultaneous equations can also be solved graphically. On a Cartesian coordinate graph, each equation would be shown as a straight line and the point at which the two cross is, in this case, $(-3, -1.5)$. See also substitution, inverse (of a matrix).

sin See sine.

sine (sin) A trigonometric function of an angle. The sine of an angle α (sinα) in a right-angled triangle is the ratio of the side opposite the angle to the hypotenuse. This definition applies only to angles between $0°$ and $90°$ (0 and $\pi/2$ radian).

More generally, in rectangular Cartesian coordinates, the y-coordinate of any point on the circumference of a circle of radius r centred at the origin is $r\sin\alpha$, where α is the angle between the x-axis and the radius to that point. In other words, the sine function depends on the vertical component of a point on a circle. Sinα is zero when α is $0°$, rises to 1 when $\alpha = 90°$ ($\pi/2$) falls again to zero when $\alpha = 180°$ (π), becomes negative and reaches -1 at $\alpha = 270°$ ($3\pi/2$), and then returns to zero at $\alpha = 360°$ (2π). This circle is repeated every complete revolution. The sine function has the following properties:

$$\sin\alpha = \sin(\alpha + 360°)$$
$$\sin\alpha = -\sin(180° + \alpha)$$
$$\sin(90° - \alpha) = \sin(90° + \alpha)$$

The sine function can also be defined as an infinite series. In the range between 1 and -1:

$$\sin x = x/1! - x^3/3! + x^5/5! - \ldots$$

See also trigonometry.

sine rule In any triangle, the ratio of the side length to the sine of the angle opposite that side is the same for all three sides. Thus, in a triangle with sides of lengths a, b, and c and angles α, β, and γ (α opposite a, β opposite b, and γ opposite c):

$$a/\sin\alpha = b/\sin\beta = c/\sin\gamma$$

sine wave The waveform resulting from plotting the sine of an angle against the angle. Any motion for which distance plotted against time gives a sine wave is a simple harmonic motion.

singular matrix A square matrix that has a determinant equal to zero and that

$$\begin{pmatrix} 2 & 1 \\ 4 & 2 \end{pmatrix}$$

$$\begin{vmatrix} 2 & 1 \\ 4 & 2 \end{vmatrix} = (2 \times 2) - (4 \times 1) = 0$$

An example of a singular 2×2 matrix

has no inverse matrix. *See also* determinant.

singular point A point on a curve $y = f(x)$ at which the derivative dy/dx has the indeterminate form $0/0$. The singular points on a curve are found by writing the derivative in the form
$$dy/dx = g(x)/h(x)$$
and then finding the values of x for which $g(x)$ and $h(x)$ are both zero.

sinh A hyperbolic sine. *See* hyperbolic functions.

sinusoidal Describing a quantity that has a waveform that is a sine wave.

SI units (*Système International d'Unités*) The internationally adopted system of units used for scientific purposes. It has seven base units (the metre, kilogram, second, kelvin, ampere, mole, and candela) and two supplementary units (the radian and steradian). Derived units are formed by multiplication and/or division of base units; a number have special names. Standard prefixes are used for multiples and submultiples of SI units. The SI system is a coherent rationalized system of units.

skewness A measure of the degree of asymmetry of a distribution. If the frequency curve has a long tail to the right and a short one to the left, it is called skewed to the right or positively skewed. If the opposite is true, the curve is skewed to the left, or negatively skewed. Skewness is measured by either Pearson's first measure of skewness (mean minus mode) divided by standard deviation, or the equivalent Pearson's second measure of skewness divided by standard deviation.

slant height 1. The length of an element of a right cone; i.e. the distance from the vertex to the directrix.
2. The altitude of the faces of a right pyramid.

slide rule A calculating device on which sliding logarithmic scales are used to multiply numbers. Most slide rules also have fixed scales showing squares, cubes, and trigonometric functions. The accuracy of the slide rule is usually to three significant figures. Slide rules have generally been replaced by electronic calculators.

sliding friction *See* friction.

slope *See* gradient.

smooth In mechanics, describing a system in which friction can be neglected in the calculations.

software The programs that can be run on a computer, together with any associated written documentation. A *software package* is a professionally written program or group of programs that is designed to perform some commonly required task, such as statistical analysis or graph plotting. The availability of software packages means that common tasks need not be programmed over and over again. *Compare* hardware. *See also* program.

solid A three-dimensional shape or object, such as a sphere or a cube.

Base and Supplementary SI Units

physical quantity	name of SI unit	symbol for unit
length	metre	m
mass	kilogram(me)	kg
time	second	s
electric current	ampere	A
thermodynamic temperature	kelvin	K
luminous intensity	candela	cd
amount of substance	mole	mol
*plane angle	radian	rad
*solid angle	steradian	sr
*supplementary units		

Derived SI Units with Special Names

physical quantity	name of SI unit	symbol for SI unit
frequency	hertz	Hz
energy	joule	J
force	newton	N
power	watt	W
pressure	pascal	Pa
electric charge	coulomb	C
electric potential difference	volt	V
electric resistance	ohm	Ω
electric conductance	siemens	S
electric capacitance	farad	F
magnetic flux	weber	Wb
inductance	henry	H
magnetic flux density	tesla	T
luminous flux	lumen	lm
illuminance (illumination)	lux	lx
absorbed dose	gray	Gy

Decimal Multiples and Submultiples to be used with SI Units

submultiple	prefix	symbol	multiple	prefix	symbol
10^{-1}	deci-	d	10^{1}	deca-	da
10^{-2}	centi-	c	10^{2}	hecto-	h
10^{-3}	milli-	m	10^{3}	kilo-	k
10^{-6}	micro-	μ	10^{6}	mega-	M
10^{-9}	nano-	n	10^{9}	giga-	G
10^{-12}	pico-	p	10^{12}	tera-	T
10^{-15}	femto-	f	10^{15}	peta-	P
10^{-18}	atto-	a	10^{18}	exa-	E

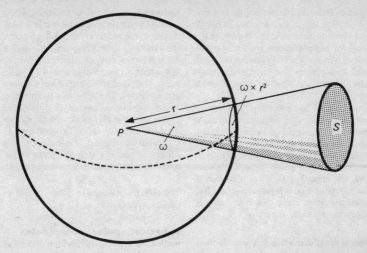

The surface S subtends a solid angle ω in steradians at point P. An area that forms part of the surface of a sphere of radius r, centre P, and subtends the same solid angle ω at P is equal to ωr^2.

solid angle Symbol: Ω The three-dimensional analogue of angle; the region subtended at a point by a surface (rather than by a line). The unit is the steradian (sr), which is defined analogously to the radian − the solid angle subtending unit area at unit distance. As the area of a sphere is $4\pi r^2$, the solid angle at its centre is 4π steradians.

solid geometry The study of geometric figures in three dimensions.

solid of revolution A solid figure that can be produced by revolution of a line or curve (the *generator*) about a fixed axis. For instance, rotating a circle about a diameter generates a sphere. Rotating a circle about an axis that does not cut the circle generates a torus.

solid-state memory See store.

solution A value of a variable that satisfies an algebraic equation. For example, the solution of $2x + 4 = 12$ is $x = 4$. An equation may have more than one solution; for example, $x^2 = 16$ has two; $x = -4$ and $x = +4$.

solution of triangles Calculating the unknown sides and angles in triangles. Because the sum of angles in a triangle is alwys $180°$, the third angle can be calculated if two are known. All the sides and angles can be calculated when two sides and the angle between them are known, but if two sides and another angle are known there are two possible solutions. Any two angles and one side are sufficient to solve a triangle. *See also* trigonometry.

source language (source program) *See* program.

space curve A curved line in a volume, defined in three-dimensional Cartesian coordinates by three functions:

$$x = f(t)$$
$$y = g(t)$$
$$z = h(t)$$

or by two equations of the form:

$$F(x,y,z) = 0$$
$$G(x,y,z) = 0$$

space–time In Newtonian (pre-relativity) physics, space and time are separate and absolute quantities; that is they are the same for all observers in any frame of reference. An event seen in one frame is also seen in the same place and at the same time by another observer in a different frame.

After Einstein had proposed his theory of relativity, Minkowski suggested that since space and time could no longer be regarded as separate continua, they should be replaced by a single continuum of four dimensions, called space–time. In space–time the history of an object's motion in the course of time is represented by a line called the *world curve*. *See also* frame of reference; relativity, theory of.

Spearman's method A way of measuring the degree of association between two rankings of n objects using two different variables x and y which give data $(x_1, y_1), \ldots, (x_n, y_n)$. The objects are ranked using first the x's and then the y's, and the difference, D, between the ranks calculated for each object.

Spearman's coefficient of rank correlation

$$\rho = 1 - (6\Sigma D^2 / [n(n^2 - 1)])$$

See also rank.

special theory *See* relativity.

spectrum The spectrum of an operator A is the set of all complex numbers λ such that the operator $A - \lambda I$, where I is the identity operator, i.e. $I(x) \equiv x$, does not have an inverse. For example, if A is a matrix, its spectrum is the set of its eigenvalues since if λ is an eigenvalue of A, $\det(A - \lambda I) = 0$ and $A - \lambda I$ is therefore not invertible.

speed Symbol: c Distance moved per unit time: $c = d/t$. Speed is a scalar quantity; the equivalent vector is velocity – a vector quantity equal to displacement per unit time. Usage can be confusing and it is common to meet the word 'velocity' where 'speed' is more correct. For instance c_0 is the speed of light in free space, not its velocity.

sphere A closed surface consisting of the locus of points in space that are at a fixed distance, the radius r, from a fixed point, the centre. A sphere is generated when a circle is turned through a complete revolution about an axis that is one of its diameters. The plane cross-sections of a sphere are all circles. The sphere is symmetrical about any plane that passes through its centre and the two mirror-image shapes on each side are called *hemispheres*. In Cartesian coordinates, the equation of a sphere of radius r with its centre at the origin is

$$x^2 + y^2 + z^2 = r^2$$

The volume of a sphere is $4\pi r^3/3$ and its surface area is $4\pi r^2$.

spherical polar coordinates A method of defining the position of a point in space by its radial distance, r, from a fixed point, the origin O, and its angular position on the surface of a sphere centred at O. The angular position is given by two angles θ and ϕ. θ is the angle that the radius vector makes with a vertical axis through O (from the south pole to the north pole). It is called the *co-latitude*. For points on the vertical axis above O, $\theta = 0$. For points lying in the 'equatorial' horizontal plane, $\theta = 90°$. For points on the vertical axis below O, $\theta = 180°$. ϕ is the angle that the radius vector makes with an axis in the equatorial plane. It is called the *azimuth*. For all points lying in the axial plane, that is, on, vertically above, or vertically below this axis, $\phi = 0$ on the positive side of O and $\phi = 180°$ on the negative side. This plane corresponds to the plane $y = 0$ in rectangular Cartesian coordinates. For points in the vertical plane at 90° to this, ($x = 0$ in rectangular Cartesian coordinates) $\phi = 90°$ in the positive half-plane and 270° in the negative half-plane. For a point P(r, θ, ϕ), the corresponding rectangular Cartesian coordinates (x, y, z) are:

$$x = r\cos\phi\sin\theta$$
$$y = r\sin\phi\sin\theta$$
$$z = r\cos\theta$$

Compare cylindrical polar coordinates. *See also* Cartesian coordinates, polar coordinates.

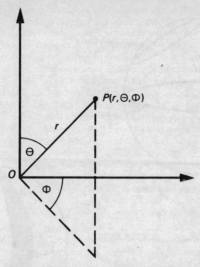

The point $P(r,\Theta,\Phi)$ in spherical polar coordinates.

spherical sector The solid generated by rotating a sector of a circle about a diameter of the circle. The volume of a spherical sector generated by a sector of altitude h (parallel to the axis of rotation) in a circle of radius r is

$$(^{2}/_{3})\pi r^2 h$$

spherical segment A solid formed by cutting in one or two parallel planes through a sphere. The volume of a spherical segment bounded by circular plane cross sections of radii r_1 and r_2 a distance h apart, is:

$$\pi h(3r_1{}^2 + 3r_2{}^2 + h^2)/6$$

If the segment is bounded by one one plane of radius r and the curved surface of the sphere, then the volume is:

$$\pi h(3r^2 + h^2)/6$$

spherical triangle A three-sided figure on the surface of a sphere, bounded by three great circles. A *right spherical triangle* has at least one right angle. A *birectangular* spherical triangle has two right angles and a *trirectangular* spherical triangle has three right angles. If one of the sides of a spherical triangle subtends an angle of 90° at the centre of the sphere, then it is called a *quadrantal spherical triangle*. An *oblique spherical triangle* has no right angles.

spherical trigonometry The study and solution of spherical triangles.

spheroid A body or curved surface that is similar to a sphere but is lengthened or shortened in one direction. *See* ellipsoid.

spiral A plane curve formed by a point winding about a fixed point at an increasing distance from it. There are many kinds of spirals, e.g. *Archimedes spiral* is given by $r = a\theta$, the *logarithmic spiral* is given by $\log(r) = a\theta$, and the *hyperbolic spiral* is given by $r\theta = a$, where in each case a is a constant, and r and θ are polar coordinates.

spur *See* square matrix.

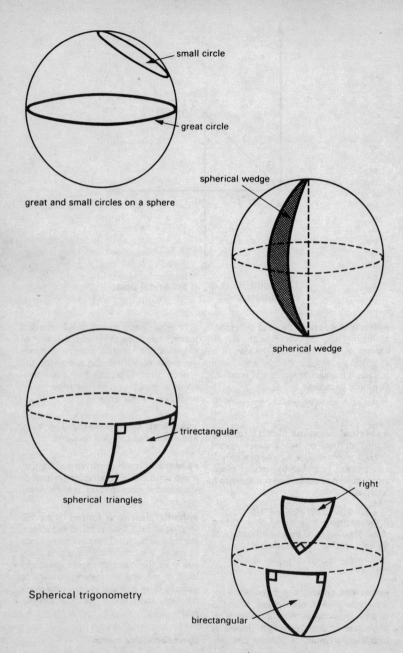

small circle

great circle

great and small circles on a sphere

spherical wedge

spherical wedge

trirectangular

spherical triangles

Spherical trigonometry

right

birectangular

$$\begin{pmatrix} 1 & 2 & 4 \\ 3 & 5 & 11 \\ 4 & 8 & 9 \end{pmatrix}$$

A 3 × 3 square matrix. The sum of the elements in the leading diagonal is $1 + 5 + 9 = 15$.

square 1. The second power of a number or variable. The square of x is $x \times x = x^2$ (x squared). The square of the square root of a number is equal to that number.
2. In geometry, a plane figure with four equal straight sides and right angles between the sides. Its area is the length of one of the sides squared. A square has four axes of symmetry – the two diagonals, which are of equal length and bisect each other perpendicularly, and the two lines joining the mid-points of opposite sides. It can be superimposed on itself after rotation through 90°.

square matrix A matrix that has the same number of rows and columns, that is, a square array of numbers. The diagonal from the top left to the bottom right of a square matrix is called the *leading diagonal* (or *principal diagonal*). The sum of the elements in this diagonal is called the *trace* (or *spur*) of the matrix.

square pyramid See pyramid.

squaring the circle The attempt to construct a square that has the same area as a particular circle, using a ruler and compasses. An exact solution is impossible because there is no exact length for the edge, which is a multiple of the transcendental number $\sqrt{\pi}$.

stability A measure of how hard it is to displace an object or system from equilibrium.

Three cases are met in statics differing in the effect on the centre of mass of a small displacement. They are:
(1) *Stable equilibrium* – the system returns to its original state when the displacing force is removed.
(2) *Unstable equilibrium* – the system moves away from the original state when displaced a small distance.
(3) *Neutral equilibrium* – when displaced a small distance, the system is at equilibrium in its new position.
An object's stability is improved by: (a) lowering the centre of mass; or (b) increasing the area of support; or by both.

stable equilibrium See stability.

standard Established as a reference.
1. Writing an equation in standard form enables comparison with other equations of the same type. For example,
$$x^2/4^2 + y^2/2^2 = 1$$
and
$$x^2/3^2 + y^2/5^2 = 1$$
are equations of hyperbolas in rectangular Cartesian coordinates, both written in standard form.
2. A standard measuring instrument is one against which other instruments are calibrated.
3. Standard form of a number. See scientific notation.

standard deviation A measure of the dispersion of a statistical sample, equal to the square root of the variance. In a sample of n observations, $x_1, x_2, x_3, \ldots x_n$, the *sample standard deviation* is:

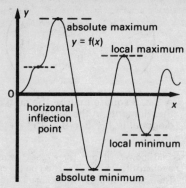

The graph of a function $y = f(x)$ showing several stationary points of different types.

$$s = \sqrt{\left[\sum_{1}^{n}(x_i - \bar{x})^2/(n-1)\right]}$$

where $\bar{x}$ is the sample mean. If the mean μ of the whole population from which the sample is taken is known, then

$$s = \sqrt{\left[\sum_{1}^{n}(x_i - \mu)^2/n\right]}$$

standard form of a number See scientific notation.

standard pressure An internationally agreed value; a barometric height of 760 mmHg at 0°C; 101 325 Pa (approximately 100 kPa).
This is sometimes called the *atmosphere* (used as a unit of pressure). The *bar*, used mainly when discussing the weather, is 100 kPa exactly. See also STP.

standard temperature An internationally agreed value for which many measurements are quoted. It is the melting temperature of water, 0°C (273.15 K). See also STP.

standing wave See stationary wave.

static friction See friction.

static pressure The pressure on a surface due to a second solid surface or to a fluid that is not flowing.

statics A branch of mechanics dealing with the forces on an object or in a system in equilibrium. In such cases there is no resultant force or torque and therefore no resultant acceleration. See also mechanics.

stationary point A point on a curved line at which the slope of the tangent to the curve is zero. All turning points (maximum points and minimum points) are stationary points. In this case the slope of the tangent passes through zero and changes its sign. Some stationary points are not turning points. In such cases, the curve levels out and then continues to increase or decrease as before. At a stationary point the derivative dy/dx of $y = f(x)$ vanishes (is zero). At a maximum, the second derivative, d^2y/dx^2, is negative; at a minimum it is positive. At a horizontal inflection point the second derivative is zero. Not all inflection points have $dy/dx = 0$; i.e. not all are stationary points.
At a stationary point on a curved surface, representing a function of two variables, $f(x,y)$, the partial derivatives $\partial f/\partial x$ and $\partial f/$

∂y are both zero. This may be a maximum point, a minimum point, or a saddle point. *See also* saddle point.

stationary wave (standing wave) The interference effect resulting from two waves of the same type moving with the same frequency through the same region. The effect is most often caused when a wave is reflected back along its own path. The resulting interference pattern is a stationary wave pattern. Here, some points always show maximum amplitude; others show minimum amplitude. They are called *antinodes* and *nodes* respectively. The distance between neighbouring node and antinode is a quarter of a wavelength.

statistical inference *See* sampling.

statistics The methods of planning experiments, obtaining data, analysing it, drawing conclusions from it, and making decisions on the basis of the analysis. In statistical inference, conclusions about a population are inferred from analysis of a sample. In descriptive statistics, data is summarized but no inferences are made.

steradian Symbol: sr The SI unit of solid angle. The surface of a sphere, for example, subtends a solid angle of 4π at its centre. The solid angle of a cone is the area intercepted by the cone on the surface of a sphere of unit radius.

stereographic projection A geometrical transformation of a sphere onto a plane. A point is taken on the surface of the sphere – the *pole* of the projection. The projection of points on the sphere onto a plane is obtained by taking straight lines from the pole through the points, and continuing them to the plane. The plane taken does not pass through the pole and is perpendicular to the diameter of the sphere through the pole. *See also* projection.

stochastic process A process that generates a series of random values of a variable and builds up a particular statistical distribution from these. For example,

the Poisson distribution can be built up by a stochastic process that starts with values taken from tables of random numbers. *See also* Poisson distribution.

store (memory) A system or device used in computing to hold information (programs and data) in such a way that any piece of information can automatically be retrieved by the computer as required. The *main store* (or *internal store*) of a computer is under the direct control of the central processor. It is the area in which programs, or parts of programs, are stored while they are being run on the computer. Data and program instructions can be extracted extremely rapidly by random access. The main store is supplemented by *backing store*, in which information can be permanently stored. The two basic forms of backing store are those in which magnetic tape is used (i.e. magnetic tape units) and those in which disks, or some other random-access device are used.

The main store is divided into a huge number of *locations*, each able to hold one unit of information, i.e. a word or a byte. The number of locations, i.e. the number of words or bytes that can be stored, gives the *capacity* of the store. Each location is identified by a serial number, known as its *address*.

There are many different ways in which memory can be classified. *Random-access memory* (RAM) and *serial-access memory* differ in the manner in which information is extracted from a store. With *volatile memory*, stored information is lost when the power supply is switched off, unlike *nonvolatile memory*. With *read-only memory* (ROM), information is stored permanently or semipermanently; it cannot be altered by programmed instructions but can in some types be changed by special techniques.

Stores may be magnetic or electronic in character. The electronic memory now widely used in main store consists of highly complex integrated circuits. This *semiconductor memory* (or *solid-state memory*) stores an immense amount of information in a very small space; items of information can be extracted at very high speed.

STP (NTP) Standard temperature and pressure. Conditions used internationally when measuring quantities that vary with both pressure and temperature (such as the density of a gas). The values are 101 325 Pa (approximately 100 kPa) and 0°C (273.15 K). *See also* standard pressure, standard temperature.

stratified random sampling *See* sampling.

Student's *t*-distribution The distribution, written t_n, of a random variable

$$t = (\bar{x} - \mu)\sqrt{n}/\sigma$$

where a random sample of size n is taken from a normal population x with mean μ and standard deviation σ. n is called the number of degrees of freedom. The mean of the distribution is 0 for $n > 1$, and the variance is $n/(n-2)$ for $n > 2$. When n is large t has an approximately standard normal distribution. The probability density function, $f(t)$, has a symmetrical graph. The values $t_n(\alpha)$ for which $P(t \leq t_n(\alpha)) = \alpha$ for various values of n are available in tables. *See also* mean, standard deviation, Student's *t*-test.

Student's *t*-test A hypothesis test for accepting or rejecting the hypothesis that the mean of a normal distribution wih unknown variance is μ_0, using a small sample. The statistic $t = (\bar{x} - \mu_0)\sqrt{n}/s$ is computed from the data $(x_1, x_2, \ldots x_n)$, where $\bar{x}$ is the sample mean, s is the sample standard deviation, and $n < 30$. If the hypothesis is true t has a t_{n-1} distribution. If t lies in the critical region $|t| > t_{n-1}(1 - \alpha/2)$ the hypothesis is rejected at significance level α. *See also* hypothesis test, Student's *t*-distribution.

subgroup A subgroup S of a group G is a subset of G that is also a group under the same law of combination of elements as G, i.e. a group whose members are members of another group. *See* group.

subject The main independent variable in an algebraic formula. For example, in the function

$$y = f(x) = 2x^2 + 3x,$$

y is the subject of the formula.

subnormal The projection on the x-axis of a line normal to a curve at point P_0 (x_0, y_0) and extending from P_0 to the x-axis. The length of the subnormal is my_0, where m is the gradient of the tangent to the curve at P_0.

subroutine (procedure) A section of a computer program that performs a task that may be required several times in different parts of the program. Instead of inserting the same sequence of instructions at a number of different points, control is transferred to the subroutine and when the task is complete it is returned to the main part of the program. *See also* routine.

subscript A small letter or number written below, and usually to the right of, a letter for various purposes, such as to identify a particular element of a set, e.g. $x_1, x_2 \in X$, to denote a constant, e.g. a_1, a_2, or to distinguish between variables, e.g. $f(x_1, x_2, \ldots, x_n)$. Double subscripts are also used, for example to write a determinant with general terms; a_{ij} denotes the element in the ith row and jth column. *See also* superscript.

subset Symbol: $\subset$ A set that forms part of another set. For example, the set of natural numbers, $N = \{1, 2, 3, 4, \ldots\}$ is a subset of the set of integers $I = \{\ldots -2, -1, 0, 1, 2, \ldots\}$, written as $N \subset I$. $\subset$ stands for the relationship of *inclusion*, and $N \subset I$ can be read: N is included in I. Another symbol used is $\supset$, which means 'contains as a subset' as in $I \supset N$. *See also* Venn diagram.

substitution A method of solving algebraic equations by replacing one variable by an equivalent in terms of another variable. For example, to solve the simultaneous equations

$$x + y = 4$$

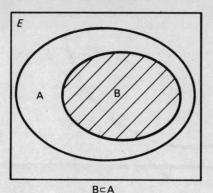

B⊂A

The set B, shaded in the Venn
diagram, is a subset of A.

and
$$2x + y = 9$$
we can first write x in terms of y, that is:
$$x = 4 - y$$
The substitution of $4 - y$ for x in the second equation gives:
$$2(4 - y) + y = 9$$
or $y = -1$, and therefore, from the first equation, $x = 5$. Another use of substitution of variables is in integration. *See also* simultaneous equations, integration by substitution.

subtangent The projection on the x-axis of the line tangent to a curve at a point $P_0 (x_0, y_0)$ and extending from P_0 to the x-axis. The length of the subtangent is y_0/m, where m is the gradient of the tangent.

subtend To lie opposite and mark out the limits of a line or angle. For example, each arc of a circle subtends a particular angle at the centre of the circle.

subtraction Symbol: − The binary operation of finding the difference between two quantities. In arithmetic, unlike addition, the subtraction of two numbers is neither commutative ($4 - 5 \neq 5 - 4$) nor associative [$2 - (3 - 4) \neq (2 - 3) - 4$]. The identity element for arithmetic subtraction is zero only when it comes on the

right-hand side ($5 - 0 = 5$, but $0 - 5 \neq 5$). In *vector subtraction*, two vectors are placed tail-to-tail forming two sides of a triangle. The length and direction of the third side gives the vector difference. Just as the sign of the difference between two numbers depends on the order of subtraction, the sense of the vector difference depends on the sense of the angle between the two vectors. *Matrix subtraction*, like matrix addition, can only be carried out between matrices with the same number of rows and columns. *Compare* addition. *See also* difference, matrix, vector difference.

subtrahend A quantity that is to be subtracted from another given quantity.

successor The successor of a member of a series is the next member of the series. In particular, the successor of an integer is the next integer, i.e. the successor of n is $n + 1$.

sufficient condition *See* condition.

sum The result obtained by adding two or more quantities together.

sum to infinity In a convergent series, the value that the sum of the first n terms,

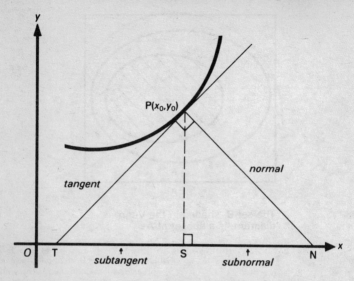

The subtangent TS and the subnormal SN of a curve at a point $P(x_0, y_0)$.

S_n approaches as n becomes infinitely large. *See* convergent series.

sup *See* supremum.

superelastic collision A collision for which the restitution coefficient is greater than one. In effect the relative velocity of the colliding objects after the interaction is greater than that before. The apparent energy gain is the result of transfer from energy within the colliding objects. For example, if a collision between two trolleys causes a compressed spring in one to be released against the other, the collision may be superelastic. *See also* restitution, coefficient of.

superscript A number or letter written above and to the right or left of a letter. A superscript usually denotes a power, e.g. x^3, or a derivative, e.g. $f^4(x) = d^4f/dx^4$, or is sometimes used in the same sense as a subscript. *See also* subscript.

supplementary angles A pair of angles that add together to make a straight line (180° or π radians). *Compare* complementary angles, conjugate angles.

supplementary units The dimensionless units – the radian and the steradian – used along with base units to form derived units. *See* SI units.

supremum (sup) The least upper bound of a set. *See* bound.

surd *See* irrational number.

surface Any locus of points extending in two dimensions. It is defined as an area. A surface may be flat (a plane surface) or curved and may be finite or infinite. For example, the plane $z = 0$ in three-dimensional Cartesian coordinates is flat and infinite; the outside of a sphere is curved and finite.

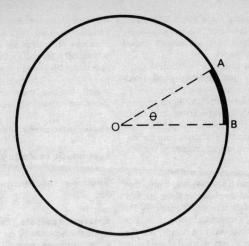

The arc AB subtends an angle θ at
the centre O of the circle.

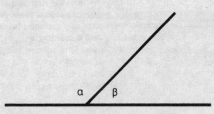

Supplementary angles: α + β = 180°

surface of revolution A surface generated by rotating a curve about an axis. For example, rotating a parabola about its axis of symmetry produces a paraboloid of revolution.

syllogism In logic, a deductive argument in which a *conclusion* is derived from two propositions, the *major premiss* and the *minor premiss*, the conclusion necessarily being true if the premisses are true. For example, 'Tim wants a car or a bicycle', 'Tim does not want a car', therefore 'Tim wants a bicycle'. A *hypothetical syllogism* is a particular type of syllogism of the form 'A implies B', 'B implies C', therefore 'A implies C'.

symbol A letter or character used to represent an object, operation, quantity, relation, or function. See the appendix for a list of mathematical symbols.

symbolic logic (formal logic) The branch of logic in which arguments, the terms used in them, the relationships between them, and the various operations that can be performed on them are all represented by symbols. The logical properties and implications of arguments can then be more easily studied strictly and formally, using algebraic techniques, proofs, and theorems in a mathematically rigorous way. It is sometimes called mathematical logic.

The simplest system of symbolic logic is *propositional logic* (sometimes called *propositional calculus*) in which letters, e.g. *P, Q, R*, etc., stand for propositions or statements, and various special symbols stand for relationships that can hold between them. *See also* biconditional, conjunction, disjunction, implication, negation, truth table.

symmetrical Denoting any figure that can be divided into two parts that are mirror images of each other. The letter A, for example, is symmetrical, and does not change when viewed in a mirror, but the letter R is not. A symmetrical plane figure has at least one line that is an axis of symmetry, which divides it into two mirror images.

syntax In logic, syntax concerns the properties of formal systems which do not depend on what the symbols actually mean. It deals with the way the symbols can be connected together and what combinations of symbols are meaningful. The syntax of a formal logical language will specify precisely and rigorously which formulae are well-formed within the language, but not their intuitive meaning. *Compare* semantics.

systematic error *See* error.

systematic sampling *See* sampling.

Système International d'Unités *See* SI units.

systems analysis The detailed analysis of the activities of an organization or system, its basic purposes, and the needs that it must satisfy, so that its performance can be improved or some other problem can be solved. The systems analyst reduces the steps required to solve a problem to a logical form. A suitable computer program can then be written to test or effect a solution, etc.

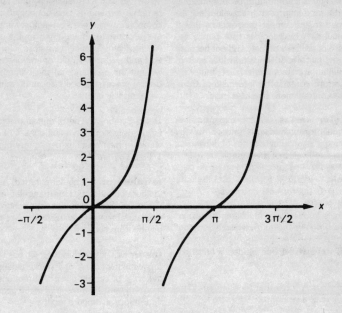

The graph of y = tan x, with x in radians.

tan *See* tangent.

tangent 1. A straight line or a plane that touches a curves or a surface without cutting through it. On a graph, the slope of the tangent to a curve is the slope of the curve at the point of contact. In Cartesian coordinates, the slope is the derivative dy/dx. If θ is the angle between the x-axis and a straight line to the point (x,y) from the origin, then the trigonometric function $\tan\theta = y/x$.

2. (tan) A trigonometric function of an angle. The tangent of an angle α in a right-angled triangle is the ratio of the lengths of the side opposite to the side adjacent. This definition applies only to angles between $0°$ and $90°$ (0 and $\pi/2$ radians). More generally, in rectangular Cartesian coordinates, with origin O, the ratio of the y-coordinate to the x-coordinate of a point P (x,y) is the tangent of the angle between the line OP and the x-axis. The tangent function, like the sine and cosine functions, is periodic, but it repeats itself every $180°$ and is not continuous. It is zero when $\alpha = 0°$, and becomes an infinitely large positive number as α approaches $90°$. At $+90°$, $\tan\alpha$ jumps from $+\infty$ to $-\infty$ and then rises to zero at $\alpha = 180°$. *See also* trigonometry.

tanh A hyperbolic tangent. *See* hyperbolic functions.

tape *See* magnetic tape, paper tape.

tautology In logic, a proposition, statement, or sentence of a form that cannot possibly be false. For example, 'if all pigs eat mice then some pigs eat mice' and 'if I am coming then I am coming' are both true regardless of whether the component propositions 'I am coming' and 'all pigs eat mice' are true or false. More strictly, a tautology is a compound proposition that is true no matter what truth values are assigned to the simple propositions that it contains. A tautology is true purely because of the laws of logic and not because of any fact about the World (the laws of thought are tautologies). A tautology therefore contains no information. *Compare* contradiction. *See also* logic.

Taylor series (Taylor expansion) A formula for expanding a function, f(x), by writing it as an infinite series of derivatives for a fixed value of the variable, $x = a$:
$f(x) = f(a) + f'(a)(x - a) +$
$f''(a)(x - a)^2/2! + f'''(a)(x - a)^3/3! + \dots$
If $a = 0$, the formula becomes:
$f(x) = f'(0) + f'(0)x = f''(0)x^2/2! + \dots$
This is known as a *Maclaurin series*, or *Maclaurin expansion*. *See also* expansion.

t-distribution *See* Student's *t*-distribution.

tension A force that tends to stretch a body (e.g. a string, rod, wire, etc.).

tensor A mathematical entity that is the general equivalent in any *n*-dimensional coordinate system, of a vector in two or three-dimensional coordinates. Tensors are used to describe how all the components of a quantity behave under certain transformations, just as a vector can describe a translation from one point to another in a plane or in space. *See also* vector.

tera- Symbol: T A prefix denoting 10^{12}. For example, 1 terawatt (TW) = 10^{12} watts (W).

terminal A point at which a user can communicate directly with a computer both for the input and output of information. It is situated outside the computer system, often at some distance from it, and is linked to it by electric cable, telephone, or some other transmission channel. A keyboard, similar to that on a typewriter, is used to feed information to the computer. The output can either be printed out or can be displayed on a screen, as with a visual display unit. An *interactive terminal* is one connected to a computer, which gives an almost immediate response to an enquiry from the user. An *intelligent terminal* can store information and perform simple operations on it without the assistance of the computer's central processor. *See also* input/output, visual display unit.

tesla Symbol: T The SI unit of magnetic flux density, equal to a flux density of one weber of magnetic flux per square metre. $1 T = 1 Wb m^{-2}$.

tetrahedron (triangular pyramid) A solid figure bounded by four triangular faces. A *regular tetrahedron* has four congruent equilateral triangles as its faces. *See also* polyhedron, pyramid.

theorem The conclusion which has been proved in the course of an argument upon the basis of certain given assumptions. A theorem must be a result of some general importance. *Compare* lemma.

theorem of parallel axes If I_0 is the moment of the inertia of an object about an axis, the moment of inertia I about a parallel axis is given by:
$$I = I_0 md^2$$
where m is the mass of the object and d is the separation of the axes.

theory of games The mathematical study of the probabilties of each outcome in games. Although there is an element of chance in who wins, there are often general rules for maximizing the chances of one particular outcome. These are calculated from the rules of the game and the number of players, using techniques of probability.

therm A unit of heat energy equal to 10^5 British thermal units (1.055 056 joules).

third-order determinant *See* determinant.

thou *See* mil.

three-dimensional Having length, breadth, and depth. A three-dimensional figure (solid) can be described in a coordinate system using three variables, for example, three-dimensional Cartesian coordinates with an x-axis, y-axis, and a z-axis. *Compare* two-dimensional.

thrust A force tending to compress a body (e.g. a rod or bar) in one direction. Thrust acts in the opposite direction to tension.

time sharing A method of operation in computer systems in which a number of jobs are apparently executed simultaneously instead of one after another (as in batch processing). This is achieved by transferring each program in turn from backing store to main store and allowing it to run for a short time. Time sharing is particularly useful for programs that are controlled by users at terminals. It allows all the users to interact with the computer apparently simultaneously, provided there are not too many of them. *Compare* batch processing.

ton 1. A unit of mass equal to 2240 pounds. It is equivalent to 1016.05 kg.
2. A unit used to express the explosive power of a nuclear weapon. It is equal to an explosion with an energy equivalent of one ton of TNT or approximately 5×10^9 joules.

tonne (metric ton) Symbol: t A unit of mass equal to 10^3 kilograms.

topologically equivalent *See* topology.

topological space A non-empty set A together with a fixed collection (T) of subsets of A satisfying:
(1) $O \in T$, $A \in T$;
(2) if $U \in T$ and $V \in T$ then $U \cap V \in T$;
(3) if $U_i \in T$, where $\{U_i\}$ is a finite or infinite collection of sets, then $\cap U_i \in T$.
The set of subsets T is called a *topology* for A, and the members of T are called the *open sets* of the topological space. *Compare* metric space. *See also* topology.

topology A branch of geometry concerned with the general properties of shapes and space. It can be thought of as the study of properties that are not changed by continuous deformations, such as stretching or twisting. A sphere and an ellipsoid are different figures in solid (Euclidean) geometry, but in topology they are considered equivalent since one can be transformed into the other by a continuous deformation. A torus, on the other hand, is not topologically equivalent to a sphere − it would not be possible to distort a sphere into a torus without breaking or joining surfaces. A torus is thus a different type of shape to a sphere. Topology studies types of shapes and their properties. A special case of this is the investigation of networks of lines and the properties of knots.

In fact Euler's study of the Königsberg bridge problem was one of the earliest results in topology. A modern example is in the analysis of electrical circuits. A circuit diagram is not an exact reproduction of the paths of the wires, but it does show the connections between different points of the circuit (i.e. it is topologically equivalent to the circuit). In printed or integrated circuits it is important to arrange connections so that they do not cross.

Topology uses methods of higher algebra including group theory and set theory. An important notion is that of sets of points in the neighbourhood of a given point (i.e. within a certain distance of the point). An *open set* is a set of points such that each point in the set has a neighbourhood containing points in the set. A topological transformation occurs when there is a one-to-one correspondence between points in one figure and points in another so that open sets in one correspond to open sets in the other. If one figure can be transformed into another by such a transformation, the sets are *topologically equivalent*.

torque Symbol: T A turning force (or moment). The torque of a force F about an axis (or point) is Fs, where s is the distance from the axis to the line of action of the force. The unit is the newton metre.

Note that the unit of *work*, also the newton metre, is called the joule. Torque is *not*, however, measured in joules. The two physical quantities are not in fact the same. Work (a scalar) is the scalar product of force and displacement. Torque is the vector product $F \times s$ and is a vector at $90°$ to the plane of the force and displacement. *See also* couple, moment.

torr A unit of pressure equal to a pressure of 101 325/760 pascals (133.322 Pa). It is equal to the mmHg.

torsional wave A wave motion in which the vibrations in the medium are rotatory simple harmonic motions around the direction of energy transfer.

torus (anchor ring) A closed curved surface with a hole in it, like a doughnut. It can be generated by rotating a circle about an axis that lies in the same plane as the circle but does not cut it. A cross-section of the torus in a plane perpendicular to the axis is two concentric circles. A cross-section in any plane that contains the axis is a pair of congruent circles at equal distances on both sides of the axis. The volume of the torus is $4\pi dr^2$ and its surface area is $3\pi^2 dr$, where r is the radius of the generating circle and d is the distance of its centre from the axis.

total derivative A derivative that can be expressed in terms of a series of partial derivatives. For example, if the function $z = f(x,y)$ is a continuous function of x and y, and both x and y are continuous functions of another variable t, then the total derivative of z with respect to t is:

$$dz/dt = (\partial z/\partial x)(dx/dt) + (\partial z/\partial y)(dy/dt)$$

See also chain rule, total differential.

total differential An infinitesimal change in a function of one or more variables. It is the sum of all the partial differentials. *See* differential.

trace *See* square matrix.

track *See* disk, drum, magnetic tape, paper tape.

trajectory 1. The path of a moving object, e.g. a projectile.
2. A curve or surface that satisfies some given set of conditions, such as passing through a given set of points, or having a given function as its gradient.

transcendental number *See* irrational number, pi.

transfinite A *transfinite number* is a cardinal or ordinal number that is not an integer. *See* cardinal number, ordinal number, aleph.
Transfinite induction is the process by which we may reason that if some proposition is true for the first element of a well-ordered set S, and that if the proposition is true for a given element it is also true for the next element, then the proposition is true for every element of S.

transformation 1. In general, any function or mapping that changes one quantity into another. *See* function.
2. The changing of an algebraic expression or equation into an equivalent with a different form. For example, the equation
$$(x - 3)^2 = 4x + 2$$
can be transformed into
$$x^2 - 10x + 7 = 0$$
3. In geometry, the changing of one shape into another by moving each point in it to a different position by a specified procedure. For example, a plane figure may be moved in relation to two rectangular axes. Another example is when a figure is enlarged. *See* translation. *See also* deformation, dilatation, enlargement, projection, rotation.

transformation of coordinates 1. Changing the position of the reference axis in a cordinate system by translation, rotation, or both, usually to simplify the equation of a curve. *See* rotation of axes, translation of axes.
2. Changing the type of coordinate system in which a geometrical figure is described; for example, from Cartesian coordinates

A transformation can be represented by a 2 × 2 matrix
A point (x,y) is transformed to a point (x',y') by multiplying the column vector of (x,y) by a matrix **M**

i.e. $\mathbf{M} \begin{pmatrix} x \\ y \end{pmatrix}$

The transformation matrices are:

reflection in x-axis $\quad \begin{pmatrix} 1 & 0 \\ 0 & -1 \end{pmatrix}$

reflection in y-axis $\quad \begin{pmatrix} -1 & 0 \\ 0 & 1 \end{pmatrix}$

enlargement scale factor k $\quad \begin{pmatrix} k & 0 \\ 0 & k \end{pmatrix}$

stretch in x-direction $\quad \begin{pmatrix} k & 0 \\ 0 & 1 \end{pmatrix}$

stretch in y-direction $\quad \begin{pmatrix} 1 & 0 \\ 0 & k \end{pmatrix}$

rotation by $\propto$
(anticlockwise positive) $\quad \begin{pmatrix} \cos \propto & -\sin \propto \\ \sin \propto & \cos \propto \end{pmatrix}$

shear in x-direction
by k $\quad \begin{pmatrix} 1 & k \\ 0 & 1 \end{pmatrix}$

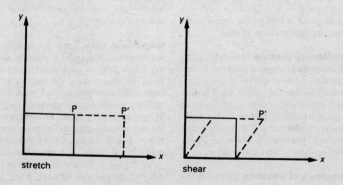

stretch

shear

187

$$A = \begin{pmatrix} a_1 & b_1 \\ a_2 & b_2 \\ a_3 & b_3 \end{pmatrix} \rightarrow \tilde{A} = \begin{pmatrix} a_1 & a_2 & a_3 \\ b_1 & b_2 & b_3 \end{pmatrix}$$

The transpose $\tilde{A}$ of a matrix A.

to polar coordinates. *See* polar coordinates.

transitive relation A relation • on a set A such that if $a \cdot b$ and $b \cdot c$ then $a \cdot c$ for all a, b, and c in A. 'Greater than', 'less than', and 'equals' are examples of transitive relations. A relation that is not transitive is an *intransitive relation*.

translation The moving of a geometrical figure so that only its position relative to fixed axes is changed, but not its orientation, size, or shape. *See also* translation of axes.

translation of axes In coordinate geometry, the shifting of the reference axes so that each axis is parallel to its original position and each point is given a new set of coordinates. For example, the origin O of a system of x- and y-axes, may be shifted to the point O', (3, 2,) in the original system. The new axes x' and y' are at $x = 3$ and $y = 2$, respectively. This is sometimes done to simplify the equation of a curve. The circle $(x - 3)^2 + (y - 2)^2 = 4$ can be described by new coordinates $x' = (x - 3)$ and $y' = (y - 3)$: $(x')^2 + (y')^2 = 4$. The origin O' is then at the centre of the circle. *See also* rotation of axes.

translatory motion (translation) Motion involving change of position; it compares with rotatory motion (rotation) and vibratory motion (vibration). Each is associated with kinetic energy. In an object undergoing translatory motion, all the points move in parallel paths. Translatory motion is usually described in terms of (linear) speed or velocity, and acceleration.

transpose of a matrix The matrix that results from interchanging the rows and columns of a given matrix. The determi-

nant of the transpose of a square matrix is equal to that of the original matrix. The transpose of a row vector is a column vector and vice versa. If two matrices A and B are comformable (can be multiplied together), then the transpose of the matrix product $AB = C$ is $\tilde{C} = (\tilde{A}\tilde{B}) = \tilde{B}\tilde{A}$. In other words, in taking the transpose of a matrix product the order must be reversed.

transverse axis *See* hyperbola.

transverse wave A wave motion in which the motion or change is perpendicular to the direction of energy transfer. Electromagnetic waves and water waves are examples of transverse waves. *Compare* longitudinal waves.

trapezium A quadrilateral with two parallel sides. It is sometimes part of the definition that the other sides are not parallel. The parallel sides are called the *bases* of the trapezoid and the perpendicular distance between the bases is called the *altitude*. The area of a trapezium is the product of the sum of the parallel side lengths and half the perpendicular distance between them.

trapezium rule A rule for finding the approximate area under a curve by dividing it into pairs of trapezium-shaped sections, forming vertical columns of equal width with bases lying on the horizontal axis. The trapezium rule is used as a method of numerical integration. For example, if the value of a function f(x) is known at $x = a$, $x = b$, and at a value mid-way between a and b, the integral is approximately:

$(h/2)\{f(a) + 2f[(a + b)/2] + f(b)\}$

h is half the distance between a and b. If this does not give a sufficiently accurate result, the area may be subdivided into 4,

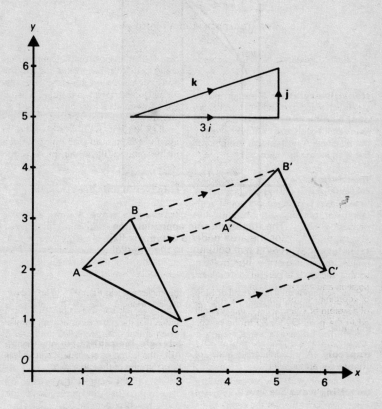

Translation

The translation of a triangle ABC
The translation vector $\mathbf{k} = 3\mathbf{i} + \mathbf{j}$

A translation by a in the x-direction and b in the y direction
transforms a point (x,y) to (x',y').

In matrix terms

$$\begin{pmatrix} x' \\ y' \end{pmatrix} = \begin{pmatrix} x \\ y \end{pmatrix} + \begin{pmatrix} a \\ b \end{pmatrix}$$

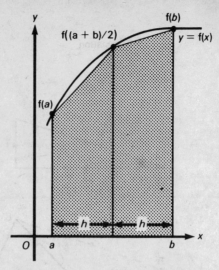

The trapezium rule approximation for the area under a curve $y = f(x)$, using two columns in the interval $x = a$ to $x = b$.

6, 8, etc., columns until further subdivision makes no significant difference to the result. *See also* numerical integration, Simpson's rule.

trapezoid A quadrilateral, none of whose sides are parallel.

travelling wave *See* wave.

triangle A plane figure with three straight sides. The area of a triangle is half the length of one side (the base) times the length of the perpendicular line (the altitude) from the base to the opposite vertex. The sum of the internal angles in a triangle is $180°$ (or π radians). In an *equilateral triangle*, all three sides have the same length and all three angles are $60°$. An *isosceles triangle* has two sides of equal length and two equal angles. A *scalene triangle* has no equal sides or angles. In a *right-angled triangle*, one angle is $90°$ (or $\pi/2$ radians), and the others therefore add up to $90°$. In an *acute angled triangle*, all

the angles are less than $90°$. In an *obtuse-angled triangle*, one of the angles is greater than $90°$.

triangle inequality For any triangle ABC, the length of one side is always less than the sum of the other two:

$$AB < BC + CA$$

triangle of forces *See* triangle of vectors.

triangle of vectors A triangle describing three coplanar vectors acting at a point with zero resultant. When drawn to scale – shown correctly in size, direction, and sense, but not in position – they form a closed triangle. Thus three forces acting on an object at equilibrium form a *triangle of forces*. Similarly a *triangle of velocities* can be constructed. *See* vector.

triangle of velocities *See* triangle of vectors.

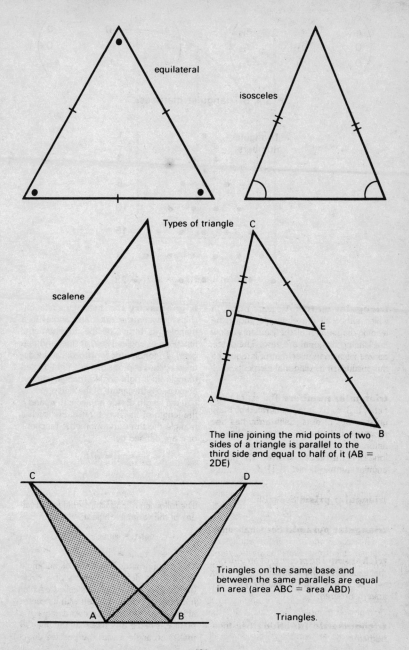

equilateral

isosceles

Types of triangle

scalene

The line joining the mid points of two sides of a triangle is parallel to the third side and equal to half of it (AB = 2DE)

Triangles on the same base and between the same parallels are equal in area (area ABC = area ABD)

Triangles.

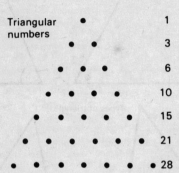

$$\begin{pmatrix} a_{11} & a_{12} & a_{13} \\ 0 & a_{22} & a_{23} \\ 0 & 0 & a_{33} \end{pmatrix} \quad \begin{pmatrix} a_{11} & 0 & 0 \\ a_{21} & a_{27} & 0 \\ a_{31} & a_{32} & a_{33} \end{pmatrix}$$

3×3 Triangular matrices.

Triangular numbers

●	1
● ●	3
● ● ●	6
● ● ● ●	10
● ● ● ● ●	15
● ● ● ● ● ●	21
● ● ● ● ● ● ●	28

triangular matrix A square matrix in which either all the elements above the leading diagonal or all the elements below the leading diagonal are zero. The determinant of a triangular matrix is equal to the product of its diagonal elements.

triangular numbers The set of numbers $\{1, 3, 6, 10, \ldots\}$ generated by triangular arrays of dots. Each array has one more row than the preceding one, the additional row having one more dot than the longest in the preceding array. The nth triangular number is $n(n + 1)/2$.

triangular prism See prism.

triangular pyramid See tetrahedron.

trichotomy The property of an ordering that exactly one of the statements $x < y$, $x = y$, $x > y$ is true for any x and y. See linearly ordered.

trigonometric functions See trigonometry.

trigonometry The study of the relationships between the sides and angles in a triangle, in terms of the trigonometric functions of angles (sine, cosine, and tangent). *Trigonometric functions*, can be defined by the ratio of sides in a right-angled triangle. In a right-angled triangle with an angle α, o is the length of the side opposite α, a the length of side adjacent to α, and h the length of the hypotenuse. For such a triangle the three trigonometric functions of α are defined by:

$$\sin\alpha = o/h$$

$$\cos\alpha = a/h$$

$$\tan\alpha = o/a$$

The following relationships hold for all values of the variable angle α:

$$\cos\alpha = \sin(\alpha + 90°)$$

$$\tan\alpha = \sin\alpha/\cos\alpha$$

The trigonometric functions of an angle can also be defined in terms of a circle (hence they are sometimes called *circular functions*). A circle is taken with its centre at the origin of Cartesian coordinates. If a point P is taken on this circle, the line OP makes an angle α with the positive direction of the x-axis. Then, the trigonometric functions are:

$$\tan\alpha = y/x$$
$$\sin\alpha = y/OP$$
$$\cos\alpha = x/OP$$

Here, (x,y) are the coordinates of the point P and $OP = \sqrt{(x^2 + y^2)}$. The signs of x and y are taken into account. For example, for an angle $\beta°$ between 90° and 180°, y will be positive and x negative. Then:

$$\tan\beta = -\tan(180 - \beta)$$
$$\sin\beta = + \sin(180 - \beta)$$
$$\cos\beta = -\cos(180 - \beta)$$

Similar relationships can be written for the trigonometric functions of angles between 180° and 270° and between 270° and 360°.

The functions secant (sec), cosecant (cosec), and cotangent (cot), which are the reciprocals of the cosine, sine, and tangent functions respectively, obey the following rules for all values of α:

$$\tan^2\alpha + 1 = \sec^2\alpha$$
$$1 + \cot^2\alpha = \csc^2\alpha$$

See also sine rule, cosine rule, addition formulae.

trillion In the US and Canada, a number represented by 1 followed by 12 zeros (10^{12}). In Britain, 1 followed by 18 zeros (10^{18}).

trinomial An algebraic expression with three variables in it. For example, $2x + 2y + z$ and $3a + b = c$ are trinomials. *Compare* binomial.

triple integral The result of integrating the same function three times. For example, if a function $f(x,y,z)$ is integrated first with respect to x, holding y and z constant, and the result is then integrated with respect to y, holding x and z constant, and the resultant double integral is then integrated with respect to z holding x and y constant, the triple integral is

$$\iiint f(x,y,z)dzdydx$$

See also double integral.

triple product A product of three vectors. *See* triple scalar product, triple vector product.

triple scalar product A product of three vectors, the result of which is a scalar, defined as:

$$\mathbf{A}.(\mathbf{B} \times \mathbf{C}) = ABC\sin\theta\cos\phi$$

where θ is the angle between $\mathbf{A}$ and the vector product ($\mathbf{B} \times \mathbf{C}$), θ being the angle between $\mathbf{B}$ and $\mathbf{C}$. The scalar triple product is equal to the volume of the parallelepiped of which $\mathbf{A}$, $\mathbf{B}$, and $\mathbf{C}$ form nonparallel edges. If $\mathbf{A}$, $\mathbf{B}$, and $\mathbf{C}$ are coplanar, their triple scalar product is zero.

triple vector product A product of three vectors, the result of which is a vector. It is a vector product of two vectors, one of which is itself a vector product. That is:

$$\mathbf{A} \times (\mathbf{B} \times \mathbf{C}) = (\mathbf{A}.\mathbf{C})\mathbf{B} - (\mathbf{A}.\mathbf{B})\mathbf{C}$$

Similarly

$$(\mathbf{A} \times \mathbf{B}) \times \mathbf{C} = (\mathbf{A}.\mathbf{C})\mathbf{B} - (\mathbf{B}.\mathbf{C})\mathbf{A}$$

These are equal only when $\mathbf{A}$, $\mathbf{B}$, and $\mathbf{C}$ are mutually perpendicular.

trirectangular Having three right angles. *See* spherical triangle.

trisection Division into three equal parts.

trivial solution A solution to an equation or set of equations that is obvious and gives no useful information about the relationships between the variables involved. For example, $x^2 + y^2 = 2x + 4y$ has a trivial solution $x = 0$; $y = 0$.

trochoid The curve described by a fixed point on the radius or extended radius of a circle as the circle rolls (in a plane) along a straight line. Let r be the radius of the circle and a the distance of the fixed point from the centre of the circle. If $a > r$ the curve is called a *prolate cycloid*, if $a < r$ the curve is called a *curtate cycloid*, and if $a = r$ the curve is a cycloid. *See also* cycloid.

truncated Describing a solid generated from a given solid by two non-parallel planes cutting the given solid.

truth table In logic, a mechanical procedure (sometimes called a matrix) that can

P Q	(P∧Q)	∨	~P
T T	T	T	F
T F	F	F	F
F T	F	T	T
F F	F	T	T

An example of a truth table

be used to define certain logical operations, and to find the truth value of complex propositions or statements containing combinations of simpler ones. A truth table lists, in rows, all the possible combinations of truth values (T = 'true', F = 'false') of a proposition or statement, and given an initial assignment of truth or falsity to the constituent parts, mechanically assigns a value to the whole. The truth-table definitions for conjunction, disjunction, negation, and implication are shown at those headwords.

An example of a truth table for a complex, or compound proposition is shown in the illustration. The assigning of values proceeds in this way: on the basis of the truth values of P and Q, the simple propositions are given truth values, written under the sign $\wedge$ (in $P \wedge Q$) and under $\sim P$. Using these, truth values can then be assigned to the whole; in the example this is in effect a complex disjunction and the values are written under the sign $\vee$.

Thus in the case where P is true and Q is false, $P \wedge Q$ will be false, $\sim P$ will be false, and therefore the whole will be false. *See also* proposition, symbolic logic.

truth value The truth or falsity of a proposition in logic. A true statement or proposition is indicated by T and a false one by F. In computer logic, the digits 1 and 0 are often used to denote truth values. *See also* truth table.

Turing machine An abstract model of a computer that consists of a control or processing unit and an infinitely long tape divided into single squares along its length. At any given time each square of

the tape is either blank or contains a single symbol from a fixed finite list of symbols $s_1, s_2, \ldots, s_n$. The Turing machine moves along the tape square by square and reads, erases, and prints symbols. At any given time the Turing machine is in one of a fixed finite number of states represented by $q_1, q_2, \ldots, q_n$. The 'program' for the machine is made up of a finite set of instructions of the form $q_i s_j s_k X q_j$, where X is either R (move to the right), L (move to the left), or N (stay in the same position). Here, q_i is the state of a machine reading s_j, which it changes to s_k, then moves left, right, or stays and completes the operation by going into state q_j. A Turing machine can be used to define computability. *See* computability.

turning point A point on the graph of a function at which the slope (gradient) of the tangent to a continuous curve changes its sign. If the slope changes from positive to negative, that is, the y-coordinate stops increasing and starts decreasing, it is a maximum point. If the slope changes from negative to positive it is a minimum point. Turning points may be local maxima and minima or absolute maxima and minima. All turning points are stationary points. At a turning point the derivative, dy/dx, of the curve $y = f(x)$ is zero. *See also* stationary point.

two-dimensional Having length and breadth but not depth. Flat shapes, such as circles, squares, and ellipses, can be described in a coordinate system using only two variables, for example, two-dimensional Cartesian coordinates with an x-axis and a y-axis. *See also* plane.

U

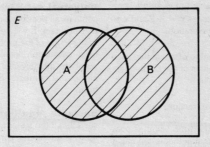

A ∪ B

The shaded area in the Venn
diagram is the union of sets A
and B.

unary operation A mathematical procedure that changes one number into another. For example, taking the square root of a number is a unary operation. *Compare* binary operation.

unbounded An *unbounded function* is a function that is not bounded − it has no bounds or limits. Intuitively this means that for any number N there is a value of the function numerically greater than N, i.e. there is a point x such that $|f(x)| > N$. For example, the function $f(x) = x$ is unbounded on the real axis, and $f(x) = 1/x$ is unbounded on the interval $0 < x \leqslant 1$.

undecidability In logic, if a formula or sentence can be proved within a given system the formula or sentence is said to be *decidable*. If every formula in a system can be proved then the system is decidable. Otherwise the system is undecidable. The process of determining which systems are decidable and which are not is an important branch of logic.

underdamping *See* damping.

uniform acceleration Constant acceleration.

uniform distribution *See* distribution function.

uniform motion A vague phrase, usually taken to mean motion at constant velocity (constant speed in a straight line).

uniform speed Constant speed.

uniform velocity Constant velocity, describing motion in a straight line with zero acceleration.

union Symbol: ∪ The combined set of all the elements of two or more sets. If A = {2, 4, 6} and B = {3, 6, 9}, then $A \cup B$ = {2, 3, 4, 6, 9}. *See also* Venn diagram.

uniqueness theorem A theorem which shows that there can only be a single entity satisfying a given condition, for example only one solution to a given equation. The proof of such a theorem usually proceeds by assuming that there exist two distinct entities satisfying the condition and showing that this leads to a contradiction.

unique solution The only possible value of a variable that can satisfy an equation. For example, $x + 2 = 4$ has the unique solution $x = 2$, but $x^2 = 4$ has no unique solution because $x = +2$ and $x = -2$ both satisfy the equation.

unit A reference value of a quantity used to express other values of the same quantity. *See also* SI units

unit fraction *See* fraction.

unit matrix (identity matrix) Symbol: I A square matrix in which the elements in the leading diagonal are all equal to one, and the other elements are zero. If a maxtrix A with m rows and n columns is multiplied by an $n \times n$ unit matrix, I, it remains unchanged, that is $IA = A$. The unit matrix is the *identity matrix* for matrix multiplication. *See also* matrix.

unit vector A vector with a magnitude of one unit. Any vector $\mathbf{r}$ can be expressed in terms of its magnitude, the scalar quantity r, and the unit vector $\mathbf{r}'$ which has the same direction as $\mathbf{r}$. The vector $\mathbf{r} = r\mathbf{r}'$, where r is the magnitude of $\mathbf{r}'$. In three-dimensional Cartesian coordinates with origin 0 unit vectors $\mathbf{i}$, $\mathbf{j}$, and $\mathbf{k}$ are used in the x-, y-, and z-directions respectively.

universal quantifier In logic, a symbol meaning 'for all' and usually written $\forall$.

For example $(\forall x)Fx$ means 'for all x, the property F is true'.

universal set Symbol: E The set that contains all possible elements. In a particular problem, E will be defined according to the scope of the problem. For example, in a calculation involving only positive numbers, the universal set, E, is the set of all positive numbers. *See also* Venn diagram.

unstable equilibrium Equilibrium such that if the system is disturbed a little, there is a tendency for it to move further from its original position rather than to return. *See* stability.

upper bound *See* bound.

upthrust An upward force on an object in a fluid. In a fluid in a gravitational field the pressure increases with depth. The pressures at different points on the object will therefore differ and the resultant is vertically upward. *See also* Archimedes' principle.

utility programs Programs that help in the general running of a computer system. They can be used, for example, to make copies of files (organized collections of data) and to transfer data from one storage device to another, as from a magnetic tape unit to a disk store. *See also* program.

V

validity In logic, a property of arguments, inferences, or deductions. An argument is valid if it is impossible that the conclusion be false while the premises are true. That is, to assert the premises and deny the conclusion would be a contradiction.

variable A changing quantity, usually denoted by a letter in algebraic equations, that might have any one of a range of possible values. Calculations can be carried out on variables because certain rules apply to all the possible values. For example, to carry out the operation of squaring all the integers between 0 and 10, and equation can be written in terms of an *integer variable n*, : $y = n^2$, with the condition that n is between 0 and 10 ($0 < n < 10$). y is called a *dependent variable* because its value depends on the value chosen for n, i.e. it can only have the values $1, 4, 9, \ldots$ etc. An *independent variable* has no such relationship with another variable. For example, if one variable, x, denotes the number of students in a school and another, y denotes the proportion of the total number of students who want to take school lunches, then x and y are independent variables and a change in either of them will not affect the other. However, their product, xy, will affect a third quantity – the number of lunches ordered. Variables may also denote quantities other than ordinary arithmetic numbers, for example, vector variables and matrix variables.

variance A measure of the dispersion of a statistical sample. In a sample of n observations $x_1, x_2, x_3, \ldots x_n$ with a sample mean $\bar{x}$, the sample variance is

$$s^2 = [(x_1 - \bar{x})^2 + (x_2 - \bar{x})^2 +$$

$$(x_3 - \bar{x})^2 + \ldots (x_n - \bar{x})^2]/(n-1)$$

See also standard deviation.

VDU *See* visual display unit.

vector A measure in which direction is important and must usually be specified. For instance, displacement is a vector quantity, whereas distance is a scalar. Weight, velocity, and magnetic field strength are other examples of vectors – they are each quoted as a number with a unit and a direction. Vectors are often denoted by printing the symbol in bold italic type **F**. *Vector algebra* treats vectors symbolically in a similar way to the algebra of scalar quantities but with different rules for addition, subtraction, multiplication, etc. Any vector can be represented in terms of component vectors. In particular, a vector in three-dimensional Cartesian coordinates can be represented in terms of three unit vector components **i, j,** and **k** directed along the x-, y-, and z-axes respectively. If P is a point with coordinates (x_1, y, z_1), then the vector OP is equivalent to $\textbf{\textit{i}}x_1 + \textbf{\textit{j}}y_1 + \textbf{\textit{k}}z_1$.
See also vector difference, vector sum, vector multiplication.

vector difference The result of subtracting two vectors. On a vector diagram, two vectors, **A** and **B**, are subtracted by drawing them tail-to-tail. The difference, **A** – **B** is the vector represented by the line from the head of **B** to the head of **A**. If **A** and **B** are parallel, the magnitude of the difference is the difference of the individual magnitudes. If they are antiparallel, it is the sum of the individual magnitudes.
The vector difference may also be calculated by taking the difference in the magnitudes of the corresponding components of each vector. For example, for two plane vectors in a Cartesian coordinate system.

$$\textbf{\textit{A}} = 4\textbf{\textit{i}} + 2\textbf{\textit{j}}$$
$$\textbf{\textit{B}} = 2\textbf{\textit{i}} + \textbf{\textit{j}}$$

where **i** and **j** are the unit vectors parallel to the x-and y-axes respectively,

$$\textbf{\textit{A}} - \textbf{\textit{B}} = 2\textbf{\textit{i}} + \textbf{\textit{j}}$$

See also vector, vector sum.

vector multiplication Multiplication of two or more vectors. This can be de-

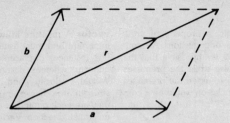

The parallelogram law r is the resultant of *a* and *b*

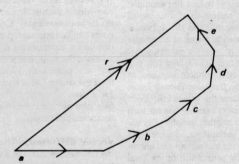

The polygon of vectors r is the resultant

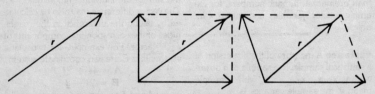

Resolution of the vector r into different pairs of components

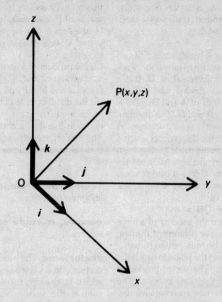

Basic vectors. The vector OP can be represented as $ix + jy + kz$

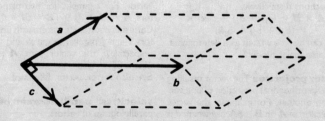

Vector product $c = a \times b$

fined in two ways according to whether the result is a vector or a scalar. *See* scalar product, vector product, triple scalar product, triple vector product.

vector product A multiplication of two vectors to give a vector. The vector product of **A** and **B** is written **A** × **B.** It is a vector of magnitude $AB\sin\theta$, where A and B are the magnitudes of **A** and **B** and θ is the angle between **A** and **B.** The direction of the vector product is at right angles to **A** and **B.** It points in the direction in which a right-hand screw would move turning from **A** towards **B.** An example of a vector product is the force **F** on a moving charge Q in a field **B** with velocity **v** (as in the motor effect). Here

$$F = QB \times v$$

Another example is the product of a force and a distance to give a moment (turning effect), which can be represented by a vector at right angles to the plane in which the turning effect acts. The vector product is sometimes called the *cross product*. The vector product is not commutative because

$$A \times B = -(B \times A)$$

It is distributive with respect to vector addition:

$$C \times (A \times B) = (C \times A) + (C \times B)$$

The magnitude of **A** × **B** is equal to the area of the parallelogram of which **A** and **B** form two non-parallel sides. In a three-dimensional Cartesian coordinate system with unit vectors **i, j,** and **k** in the x, y, and z directions respectively,

$$A \times B = (a_1 i + a_2 j + a_3 k) \times$$
$$(b_1 i + b_2 j + b_3 k)$$

This can also be written as a determinant. *See also* scalar product.

vector projection The vector resulting from an orthogonal projection of one vector on another. For example, the vector projection of **A** on **B** is **b**$A\cos\theta$ where θ is the smaller angle between **A** and **B,** and **b** is the unit vector in the direction of **B.** *Compare* scalar projection.

vector space A *vector space* V over a *scalar field* F is a set of elements x, y, ..., called *vectors*, together with two algebraic operations called vector addition and multiplication of vectors by scalars, i.e. by elements of F, such that:

(1) the sum of two elements is written $x + y$, and V is an Abelian group with respect to addition;

(2) the product of a vector, x, and a scalar, a, is written ax and $a(bx) = (ab)x$, $1x = x$ for all a and b in F and all x in V;

(3) the distributive laws hold, i.e. $a(x + y) = ax + ay$, $(a + b)x = ax + bx$ for all a and b in F and all x and y in V.

The scalars may be real numbers, complex numbers, or elements of some other field.

vectors, parallelogram of *See* parallelogram of vectors.

vectors, triangle of *See* triangle of vectors.

vector sum The result of adding two vectors. On a vector diagram, vectors are added by drawing them head to tail. The sum is the vector represented by the straight line from the tail of the first to the head of the last. If they are parallel, the magnitude of the sum is the sum of the magnitudes of the individual vectors. If two vectors are anti-parallel, then the magnitude of the sum is the difference of individual magnitudes.

The vector sum may also be calculated by summing the magnitudes of the corresponding components of each individual vector. For example, for two plane vectors, **A** = $2i + 3j$ and **B** = $6i + 4j$, in a Cartesian coordinate system with unit vectors **i** and **j** parallel to the x- and y-axes respectively, the vector sum, **A** + **B**, equals $8i + 4j$.

See also vector, vector difference.

velocities, parallelogram of *See* parallelogram of vectors.

velocities, triangle of *See* triangle of vectors.

velocity Symbol: v Displacement per unit time. The unit is the metre per second $(m\ s^{-1})$. Velocity is a vector quantity,

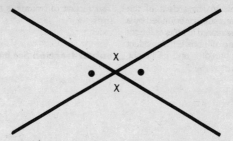

Vertically opposite angles formed at
the intersection of two straight lines.

speed being the scalar form. If velocity is
constant, it is given by the slope of a posi-
tion/time graph, and by the displacement
divided by the time taken. If it is not constant, the mean value is obtained. If x is
the displacement, the instantaneous velocity is given by

$$v = dx/dt$$

See also equations of motion.

velocity ratio *See* distance ratio.

Venn diagram A diagram used to show
the relationships between sets. The universal set, E, is shown as a rectangle. Inside
this, other sets are shown as circles. Intersecting or overlapping circles are intersecting sets. Separate circles are sets that have
no intersection. A circle inside another is a
subset. A group of elements, or a subset,
defined by any of these relationships may
be indicated by a shaded area in the diagram. *See also* set.

verifier *See* card.

vertex 1. A point at which lines or planes
meet in a figure; for example, the top point
of a cone or pyramid, or a corner of a
polygon or polyhedron.
2. One of the two points at which an axis
of a conic cuts the conic. *See* ellipse, hyperbola, parabola.

vertically opposite angles One of
the two pairs of equal angles formed when
two straight lines cross each other.

vibration (oscillation) Any regularly repeated to-and-fro motion or change. Examples are the swing of a pendulum, the
vibration of a sound source, and the
change with time of the electric and magnetic fields in an electromagnetic wave.

virtual work The work done if a system
is displaced infinitesimally from its position. The virtual work is zero if the system
is in equilibrium.

visual display unit (VDU) A computer terminal with which the user can communicate with the computer by means of a
keyboard similar to that of a typewriter;
this input and also the output from the
computer appears on a cathode-ray
screen. A VDU can operate both as in input and output device. The information
displayed is in the form of words, numbers, etc. A *graphics display* is a similar
device in which information appears as
graphs or other drawings as well as text.

volatile memory (volatile store) *See*
store.

volt Symbol: V The SI unit of electrical
potential, potential difference, and e.m.f.,
defined as the potential difference between two points in a circuit between
which a constant current of one ampere
flows when the power dissipated is one
watt. One volt is one joule per coulomb (1
$V = 1 \text{ J C}^{-1}$).

volume Symbol: *V* The extent of the space occupied by a solid or bounded by a closed surface, measured in units of length cubed. The volume of a box is the product of its length, its breadth, and its height.

The SI unit of volume is the cubic metre (m^3).

vulgar fraction *See* fraction.

W

watt Symbol: W The SI unit of power, defined as a power of one joule per second. $1\,W = 1\,J\,s^{-1}$.

wave A method of energy transfer involving some form of vibration. For instance, waves on the surface of a liquid or along a stretched string involve regular to-and-fro motion of particles about a mean position. Sound waves carry energy by alternate compressions and rarefactions of air (or other media). In electromagnetic waves, electric and magnetic fields vary at right angles to the direction of propagation of the wave. At any particular instance, a graph of displacement against distance is a regular repeating curve – the *waveform* or wave profile of the wave. In a *travelling* (or *progressive*) wave the whole periodic displacement moves through the medium. At any point in the medium the disturbance is changing with time. Under certain conditions a stationary (or standing) wave can be produced in which the disturbance does not change with time.

For the simple case of a plane progressive wave the displacement at a point can be represented by an equation:
$$y = a\sin2\pi(ft - x/\lambda)$$
where a is the amplitude, f the frequency, x the distance from the origin, and λ the wavelength. Other relationships are:
$$y = a\sin2\pi(vt - x)/\lambda$$
where v is the speed, and
$$y = a\sin2\pi(t/T - x/\lambda)$$
where T is the period. Note that if the minus sign is replaced by a plus sign in the above equations it implies a similar wave moving in the opposite direction. For a stationary wave resulting from two waves in opposite directions, the displacement is given by:
$$Y = 2a\cos2\pi x/\lambda$$
See also longitudinal wave, stationary wave, transverse wave, phase.

wave equation A second order partial differential equation that describes wave motion. The equation

$$\partial^2 u/\partial x^2 = (1/c^2)\partial^2 u/\partial t^2$$

might represent, for example, the vertical displacement u of the water surface as a plane wave of velocity c passes along the surface, the horizontal position and the time being given by x and t respectively. The general solution of this one-dimensional wave equation is a periodic function of x and t.

waveform *See* wave.

wavefront A continuous surface associated with a wave radiation, in which all the vibrations concerned are in phase. A parallel beam has plane wavefronts; the output of a point source has spherical wavefronts.

wavelength Symbol: λ The distance between the ends of one complete cycle of a wave. Wavelength relates to the wave speed (c) and its frequency (v) thus:
$$c = v\lambda$$

wave motion Any form of energy transfer that may be described as a wave rather than as a stream of particles. The term is also sometimes used to mean any harmonic motion.

wave number Symbol: σ The reciprocal of the wavelength of a wave. It is the number of wave cycles in unit distance, and is often used in spectroscopy. The unit is the metre^{-1} (m^{-1}). The circular wave number (Symbol: k) is given by
$$k = 2\pi\sigma$$

weber Symbol: Wb The SI unit of magnetic flux, equal to the magnetic flux that, linking a circuit of one turn, produces an e.m.f. of one volt when reduced to zero at a uniform rate in one second. $1\,Wb = 1\,V\,s$.

weight Symbol: W The force by which a mass is attracted to another, such as the Earth. It is proportional to the body's mass (m), the constant of proportionality being the gravitational field strength (i.e. the acceleration of free fall). Thus $W = mg$, where g is the acceleration of free fall. The mass of a body is normally constant, but its weight varies with position (because it depends on g).

Although mass and weight are often used interchangeably in everyday language, they are different in scientific language and must not be confused.

weighted mean *See* mean.

weightlessness An apparent loss of weight experienced by an object in free fall. Thus for a person in an orbiting spacecraft, the weight in the Earth's frame of reference is the centripetal force necessary to maintain the circular orbit. In the frame of reference of the spacecraft the person feels that he has no weight.

wheel and axle A simple machine consisting of a wheel on an axle that has a rope around it. An effort applied to the wheel is transmitted to a load exerted at the axle rope. The force ratio (mechanical advantage) is equal to r_W/r_A where r_W is the radius of the wheel and r_A that of the axle. *See* machine.

whole numbers Symbol: W The set of integers $\{1,2,3, \dots\}$, excluding zero.

word The basic unit in which information is stored and manipulated in a computer. Each word consists usually of a fixed number of bits. This number, known as the *word length*, varies with the type of computer and may be as few as eight or as many as 60. Each word is given a unique address in store. A word may represent an instruction to the computer or a piece of data. An instruction word is coded to give the operation to be performed and the address or addresses of the data on which the operation is to be performed. *See also* bit, byte.

work Symbol: W The work done by a force is the product of the force and the distance moved in the same direction:

work = force × displacement

Work is in fact a process of energy transfer and, like energy, is measured in joules. If the directions of force (F) and motion are not the same the component of the force in the direction of the motion is used.

$$W = Fs\cos\theta$$

where s is displacement and θ the angle between the directions of force and motion. Work is the scalar product of force and displacement.

world curve *See* space–time.

Y, Z

yard A unit of length now defined as 0.9144 metre.

yield The income produced by a stock or share expressed as a percentage of its market price.

zero (0) The number that when added to another number gives a sum equal to that other number. It is included in the set of integers but not in the set of whole numbers. The product of any number and zero is zero. Zero is the identity element for addition.

zero matrix *See* null matrix.

zone A part of a sphere produced by two parallel planes cutting the sphere.

Zorn's lemma If a set S is partially ordered and each linearly ordered subset has an upper bound in S, then S contains at least one maximal element, i.e. an element x such that there is no y in S for $x < y$.

APPENDIX

Contents

Symbols and Notation

Arithmetic and algebra

equal to	$=$
not equal to	$\neq$
identity	$\equiv$
approximately equal to	$\approx$
approaches	$\rightarrow$
proportional to	$\propto$
less than	$<$
greater than	$>$
less than or equal to	$\leqslant$
greater than or equal to	$\geqslant$
much less than	$\ll$
much greater than	$\gg$
plus, positive	$+$
minus, negative	$-$
multiplication	$a \times b$
	$a.b$
division	$a \div b$
	a/b
magnitude of a	$|a|$
factorial a	$a!$
logarithm (to base b)	$\log_b a$
common logarithm	$\log_{10} a$
natural logarithm	$\log_e a$
	$\ln a$
summation	Σ
continued product	Π

Geometry and trigonometry

angle	$\angle$
triangle	$\triangle$
square	$\square$
circle	$\bigcirc$
parallel to	$\parallel$
perpendicular to	$\perp$
congruent to	$\equiv$
similar to	$\sim$
sine	sin
cosine	cos
tangent	tan
contangent	cot, ctn
secant	sec
cosecant	cosec, csc
inverse sine	$\sin^{-1}$, arc sin
etc.	etc.
Cartesian coordinates	(x, y, z)
spherical coordinates	(r, θ, ϕ)
cylindrical coordinates	(r, θ, z)
direction numbers	l, m, n
or cosines	

Sets and logic

implies that	$\Rightarrow$
is implied by	$\Leftarrow$
implies and is implied by (if and only if)	$\Leftrightarrow$
set $a, b, c, \ldots$	$\{a, b, c, \ldots \}$
is an element of	$\in$
is not an element of	$\notin$
such that	:
number of elements in the set S	$n(S)$
universal set	E or ε
empty set	ϕ
complement of S	S'
union	$\cup$
intersection	$\cap$
is a subset of	$\subset$
corresponds one-to-one with	$\leftrightarrow$
x is mapped onto y	$x \rightarrow y$
the set of natural numbers	N
the set of integers	Z
the set of rational numbers	Q
the set of real numbers	R
the set of complex numbers	C
conjunction	$\wedge$
disjunction	$\vee$
negation	$\sim p$ or $\neg p$
implication	$\rightarrow$ or $\supset$
biconditional (equivalence)	$\equiv$ or $\leftrightarrow$

Symbols and notation continued

Calculus

increment of x	$\Delta x,\ dx$
limit of function of x as $x \rightarrow a$	$\lim_{x \rightarrow a} f(x)$
derivative of $f(x)$	$df(x)/dx,\ f'(x)$
second derivative etc.	$d^2f(x)/dx,\ f''(x)$
indefinite integral	$\int f(x)\ dx$
definite integral with limits a and b	$\int_b^a f(x)\ dx$
partial derivative with respect to x	$\partial f(x,\ y)/\partial x$

Symbols for Physical Quantities

Quantity	Symbol
acceleration	a
angle	α, etc.
angular acceleration	α
angular frequency $2\pi f$	ω
angular momentum	L
angular velocity	ω
area	A
breadth	b
circular wavenumber	k
density	ρ
diameter	d
distance	s, L
energy	W, E
force	F
frequency	$f,$
height	h
kinetic energy	E_k, T
length	l
mass	m
moment of force	M
moment of inertia	I
momentum	p
period	T
potential energy	E_p, V
power	P
pressure	p
radius	r
reduced mass $m_1 m_2/(m_1 + m_2)$	μ
relative density	d
solid angle	Ω, ω
thickness	d
time	t
torque	T
velocity	v
volume	V
wavelength	λ
wavenumber	σ
weight	W
work	W, E

Imperial Units

SI units are now used for all scientific purposes and metric units are coming into general use for commercial purposes. The tables given here show the Imperial system of 'weights and measures'. Some, such as the mile, gallon and pint, are in every-day use. Others — the league or the peck, for example — are only of historical interest. To convert between Imperial and SI units, see the conversion factors later in this appendix.

Linear measure

1 mil	0.001 inch
1 inch	1000 mils
12 inches	1 foot
3 feet	1 yard
5.5 yards	1 rod (pole, or perch)
4 rods	1 chain
10 chains	1 furlong
8 furlongs	1 statute mile
3 miles	1 league

Nautical linear measure

6 feet	1 fathom
100 fathoms	1 cable's length
10 cable's length	1 international nautical mile

Surveyors measure

7.92 inches	1 link
100 links	1 chain
80 chains	1 mile

Square measure

144 square inches	1 square foot
9 square feet	1 square yard
4840 square yards	1 acre
640 acres	1 square mile

Cubic measure
(see also liquid and dry measure)

1728 cubic inches	1 cubic foot
27 cubic feet	1 cubic yard

Liquid measure

60 minims	1 fluid drachm
8 fluid drachms	1 fluid ounce
5 fluid ounces	1 gill
20 fluid ounces	1 UK pint
16 fluid ounces	1 US pint
2 pints	1 quart
4 quarts	1 gallon
31.5 gallons	1 barrel
2 barrels	1 hogshead

Note there is a difference between the UK and US systems: 1 gallon (US) is 0.83268 gallon (UK)

Dry measure

2 pints	1 quart
8 quarts	1 peck
4 pecks	1 bushel

Units of mass (or weight)

Three systems have been in use: the avoirdupois system, used for general purposes; the troy system, used for precious metals or gemstones; and the apothecaries' system, used formerly in pharmacy. In each system the grain has the same value (0.0648 gram). Note that in the avoirdupois system there is a difference in usage between the UK and the US. In the UK the long hundredweight (112lb) and long ton (2240lb) are used. In the US the short hundredweight (100lb) and short ton (2000lb) are used.

Avoirdupois system

$27\frac{11}{32}$ grains	1 dram
16 drams	1 ounce
16 ounces	1 pound
14 pounds	1 stone
2 stones	1 quarter
4 quarters	1 long hundredweight
100 pounds	1 short hundredweight
20 hundredweights	1 ton

Troy system

24 grains	1 pennyweight
20 pennyweights	1 ounce
12 ounces	1 pound

Apothecaries' system

20 grains	1 scruple
3 scruples	1 drachm
8 drachms	1 ounce
12 ounces	1 pound

Conversion Factors

Length

To convert	into	multiply by
inches	metres	0.0254
feet	metres	0.3048
yards	metres	0.9144
miles	kilometres	1.60934
nautical miles	kilometres	1.85200
nautical miles	miles	1.15078
kilometres	miles	0.621371
kilometres	nautical miles	0.539957
metres	inches	39.3701
metres	feet	3.28084
metres	yards	1.09361

Area

To convert	into	multiply by
square inches	square centimetres	6.4516
square inches	square metres	6.4516×10^{-4}
square feet	square metres	9.2903×10^{-2}
square yards	square metres	0.836127
square miles	square kilometres	2.58999
square miles	acres	640
acres	square metres	4046.86
acres	square miles	1.5625×10^{-3}
square centimetres	square inches	0.155
square metres	square feet	10.7639
square metres	square yards	1.19599
square metres	acres	2.47105×10^{-4}
square metres	square miles	3.86019×10^{-7}
square kilometres	square miles	0.386019

Volume

To convert	into	multiply by
cubic inches	litres	1.63871×10^{-2}
cubic inches	cubic metres	1.63871×10^{-5}
cubic feet	litres	28.3168
cubic feet	cubic metres	0.0283168
cubic yard	cubic metres	0.764555
gallon (British)	litres	4.54609
gallon (British)	cubic metres	4.54609×10^{-3}
gallon (USA)	gallon (British)	0.83268

Mass

To convert	into	multiply by
pounds	kilograms	0.453592
pounds	tonnes	4.53592×10^{-4}
stones	kilograms	6.350293
stones	tonnes	6350.293
hundredweight	kilograms	50.802345
hundredweight	tonnes	0.050802345
tons	kilograms	1016.047
tons	tonnes	1.016047
kilograms	pounds	2.204623
kilograms	stones	0.157473
kilograms	hundredweights	0.01968413
kilograms	tons	9.842065×10^{-4}
tonnes	pounds	2204.623
tonnes	stones	157.473
tonnes	hundredweights	19.68413
tonnes	tons	0.9842065

Force

To convert	into	multiply by
pounds force	newtons	4.44822
pounds force	kilograms force	0.453592
pounds force	dynes	444822
pounds force	poundals	32.174
poundals	newtons	0.138255
poundals	kilograms force	0.0140981
poundals	dynes	13825.5
poundals	pounds force	0.031081
dynes	newtons	10^{-5}
dynes	kilograms force	1.01972×10^{-6}
dynes	pounds force	2.24809×10^{-6}
dynes	poundals	7.2330×10^{-5}
kilograms force	newtons	9.80665
kilograms force	dynes	980665
kilograms force	pounds force	2.20462
kilograms force	poundals	70.9316
newtons	kilograms	0.101972
newtons	dynes	100000
newtons	pounds force	0.224809
newtons	poundals	7.23300

Work and energy

To convert	into	multiply by
British Thermal Units	joules	1055.06
British Thermal Units	calories	251.997
British Thermal Units	kilowatt hours	2.93071×10^{-4}
kilowatt hours	joules	3600000
kilowatt hours	calories	859845
kilowatt hours	British Thermal Units	3412.14
calories	joules	4.1868
calories	kilowatt hours	1.16300×10^{-6}
calories	British Thermal Units	3.96831×10^{-3}
joules	calories	0.238846
joules	kilowatt hours	2.7777×10^{-7}
joules	British Thermal Units	9.47813×10^{-4}
joules	electron volts	6.2418×10^{18}
joules	ergs	10^7
electron volts	joules	1.6021×10^{-19}
ergs	joules	10^{-7}

Pressure

To convert	into	multiply by
atmospheres	pascals*	101325
atmospheres	kilograms per square metre	10332
atmospheres	pounds per square inch	14.6959
atmospheres	bars	1.01325
atmospheres	torrs	760
bars	pascals	100000
pounds per square inch	pascals	6894.76
pounds per square inch	kilograms per square metre	703.068
pounds per square inch	atmospheres	0.068046
kilograms per square metre	pascals	9.80661
kilograms per square metre	pounds per square inch	1.42234×10^{-3}
kilograms per square metre	atmospheres	9.67841×10^{-5}
pascals	kilograms per square metre	0.101972
pascals	pounds per square inch	1.45038×10^{-4}
pascals	atmospheres	9.86923×10^{-6}

* 1 pascal = 1 newton per square metre

Density

To convert	into	multiply by
pounds per cubic foot	kilograms per cubic metre	16.018463
pounds per cubic foot	grams per cubic centimetre	0.016018463
pounds per cubic inch	kilograms per cubic metre	27679.9
pounds per cubic inch	grams per cubic centimetre	27.6799
kilograms per cubic metre	pounds per cubic foot	0.062428
kilograms per cubic metre	pounds per cubic inch	3.61273×10^{-5}
grams per cubic centimetre	pounds per cubic foot	62.428
grams per cubic centimetre	pounds per cubic inch	0.0361273

Mensuration

Figure		Area
triangle	sides a, b, angle A	½ bc sin$A$
square	side a	a^2
rectangle	sides a and b	$a \times b$
kite	diagonals c and d	½ $c \times d$
parallelogram	sides a and b distances c and d apart	$a \times c = b \times d$
circle	radius r perimeter $2\pi r$	πr^2
ellipse	axes a and b perimeter $2\pi\sqrt{[(a^2+b^2)/2]}$	πab

		Surface area	Volume
cylinder	radius r height h	$2\pi r(h+r)$	$\pi r^2 h$
cone	base radius r slant height l height h	πrl	$\pi r^2 h/3$
sphere	radius r	$4\pi r^2$	$4\pi r^3/3$

Expansions

$\sin x \quad = x/1! - x^3/3! + x^5/5! - x^7/7! + \ldots$

$\cos x \quad = 1 - x^2/2! + x^4/4! - x^6/6! + \ldots$

$e^x \quad = 1 + x/1! + x^2/2! + x^3/3! + \ldots$

$\sinh x \quad = x + x^3/3! + x^5/5! + x^7/7! + \ldots$

$\cosh x \quad = 1 + x^2/2! + x^4/4! + x^6/6! + \ldots$

$\ln(1 + x) = x - x^2/2 + x^3/3 - x^4/4 + \ldots \text{ for } |x| < 1$

$(1 + x)^n = 1 + nx + n(n - 1)x^2/2! + \ldots + (\ ) x^r + \ldots \text{ for } |x| < 1$

$f(a + x) = f(a) + xf'(a) + (x^2/2!)f''(a) + (x^3/3!)f''' (a) + \ldots$

$f(x) \quad = f(0) + xf' (0) + (x^2/2!)f'' (0) + (x^3/3!)f''' (0) + \ldots$

Derivatives

x is a variable, u is a function of x, a and n are constants.

Function	Derivative
x	1
ax	a
ax^n	anx^{n-1}
e^{ax}	ae^{ax}
$\log_e x$	$1/x$
$\log_a x$	$(1/x)\log_e a$
$\cos x$	$-\sin x$
$\sin x$	$\cos x$
$\tan x$	$\sec^2 x$
$\cot x$	$-\csc^2 x$

Derivatives continued

Function	Derivative
$\sec x$	$\tan x.\ \sec x$
$\operatorname{cosec} x$	$-\cot x.\ \operatorname{cosec} x$
$\cos u$	$-\sin u.\ (du/dx)$
$\sin u$	$\cos u.\ (du/dx)$
$\tan u$	$\sec^2 u.\ (du/dx)$
$\log_e u$	$(1/u)\ (du/dx)$
$\sin^{-1}(x/a)$	$1/\sqrt{(a^2 - x^2)}$
$\cos^{-1}(x/a)$	$-1/\sqrt{(a^2 - x^2)}$
$\tan^{-1}(x/a)$	$a/(a^2 + x^2)$

Integrals

x is a variable. a and n are constants. The constant of integration C should be added to each integral.

Function	Integral
x^n	$x^{n+1}/(n+1)$
$1/x$	$\log_e x$
e^{ax}	e^{ax}/a
$\log_e ax$	$x\log_e ax - x$
$\cos x$	$\sin x$
$\sin x$	$-\cos x$
$\tan x$	$\log_e(\cos x)$
$\cot x$	$\log_e(\sin x)$
$\sec x$	$\log_e(\sec x + \tan x)$
$\operatorname{cosec} x$	$\log_e(\operatorname{cosec} x - \cot x)$
$1/\sqrt{(a^2 - x^2)}$	$\sin^{-1}(x/a)$
$-1/\sqrt{(a^2 - x^2)}$	$\cos^{-1}(x/a)$

Powers and Roots

n	n^2	n^3	$\sqrt{n}$	$\sqrt[3]{n}$
1	1	1	1.000	1.000
2	4	8	1.414	1.260
3	9	27	1.732	1.442
4	16	64	2.000	1.587
5	25	125	2.236	1.710
6	36	216	2.449	1.817
7	49	343	2.646	1.913
8	64	512	2.828	2.000
9	81	729	3.000	2.080
10	100	1 000	3.162	2.154
11	121	1 331	3.317	2.224
12	144	1 728	3.464	2.289
13	169	2 197	3.606	2.351
14	196	2 744	3.742	2.410
15	225	3 375	3.873	2.466
16	256	4 096	4.000	2.520
17	289	4 913	4.123	2.571
18	324	5 832	4.243	2.621
19	361	6 859	4.359	2.668
20	400	8 000	4.472	2.714
21	441	9 261	4.583	2.759
22	484	10 648	4.690	2.802
23	529	12 167	4.796	2.844
24	576	13 824	4.899	2.884
25	625	15 625	5.000	2.924
26	676	17 576	5.099	2.962
27	729	19 683	5.196	3.000
28	784	21 952	5.292	3.037
29	841	24 389	5.385	3.072
30	900	27 000	5.477	3.107

Powers and roots continued

n	n^2	n^3	$\sqrt{n}$	$\sqrt[3]{n}$
31	961	29 791	5.568	3.141
32	1 024	32 768	5.657	3.175
33	1 089	35 937	5.745	3.208
34	1 156	39 304	5.831	3.240
35	1 225	42 875	5.916	3.271
36	1 296	46 656	6.000	3.302
37	1 369	50 653	6.083	3.332
38	1 444	54 872	6.164	3.362
39	1 521	59 319	6.245	3.391
40	1 600	64 000	6.325	3.420
41	1 681	68 921	6.403	3.448
42	1 764	74 088	6.481	3.476
43	1 849	79 507	6.557	3.503
44	1 936	85 184	6.633	3.530
45	2 025	91 125	6.708	3.557
46	2 116	97 336	6.782	3.583
47	2 209	103 823	6.856	3.609
48	2 304	110 592	6.928	3.634
49	2 401	117 649	7.000	3.659
50	2 500	125 000	7.071	3.684
51	2 601	132 651	7.141	3.708
52	2 704	140 608	7.211	3.733
53	2 809	148 877	7.280	3.756
54	2 916	157 464	7.348	3.780
55	3 025	166 375	7.416	3.803
56	3 136	175 616	7.483	3.826
57	3 249	185 193	7.550	3.849
58	3 364	195 112	7.616	3.871
59	3 481	205 379	7.681	3.893
60	3 600	216 000	7.746	3.915
61	3 721	226 981	7.810	3.936
62	3 844	238 328	7.874	3.958
63	3 969	250 047	7.937	3.979
64	4 096	262 144	8.000	4.000
65	4 225	274 625	8.062	4.021

n	n^2	n^3	$\sqrt{n}$	$\sqrt[3]{n}$
66	4 356	287 496	8.124	4.041
67	4 489	300 763	8.185	4.062
68	4 624	314 432	8.246	4.082
69	4 761	328 509	8.307	4.102
70	4 900	343 000	8.367	4.121
71	5 041	357 911	8.426	4.141
72	5 184	373 248	8.485	4.160
73	5 329	389 017	8.544	4.179
74	5 476	405 224	8.602	4.198
75	5 625	421 875	8.660	4.217
76	5 776	438 976	8.718	4.236
77	5 929	456 533	8.775	4.254
78	6 084	474 552	8.832	4.273
79	6 241	493 039	8.888	4.291
80	6 400	512 000	8.944	4.309
81	6 561	531 441	9.000	4.327
82	6 724	551 368	9.055	4.344
83	6 889	571 787	9.110	4.362
84	7 056	592 704	9.165	4.380
85	7 225	614 125	9.220	4.397
86	7 396	636 056	9.274	4.414
87	7 569	658 503	9.327	4.431
88	7 744	681 472	9.381	4.448
89	7 921	704 969	9.434	4.465
90	8 100	729 000	9.487	4.481
91	8 281	753 571	9.539	4.498
92	8 464	778 688	9.592	4.514
93	8 649	804 357	9.644	4.531
94	8 836	830 584	9.695	4.547
95	9 025	857 375	9.747	4.563
96	9 216	884 736	9.798	4.579
97	9 409	912 673	9.849	4.595
98	9 604	941 192	9.899	4.610
99	9 801	970 299	9.950	4.626
100	10 000	1 000 000	10.000	4.642

Important Constants

speed of light	$2.997\,925 \times 10^{8}$	m s^{-1}
Planck constant	$6.626\,196 \times 10^{-34}$	J s
Boltzmann constant	$1.380\,622 \times 10^{-23}$	JK^{-1}
Avogadro constant	$6.022\,169 \times 10^{23}$	mol^{-1}
mass of proton	$1.672\,614 \times 10^{-27}$	kg
mass of neutron	$1.674\,920 \times 10^{-27}$	kg
mass of electron	$9.109\,558 \times 10^{-31}$	kg
charge of proton or electron	$\pm 1.602\,191\,\Delta 7 \times 10^{-19}$	C
specific charge of electron	$-1.758\,796 \times 10^{11}$	C kg^{-1}
molar volume at s.t.p.	$2.241\,36 \times 10^{-2}$	m^3 mol^{-1}
Faraday constant	$9.648\,670 \times 10^{4}$	C mol^{-1}
triple point of water	273.16	K
absolute zero	-273.15	°C
permittivity of vacuum	$8.854\,185\,3 \times 10^{-12}$	F m^{-1}
permeability of vacuum	$4\pi \times 10^{-7}$	H m^{-1}
Stefan constant	$5.669\,61 \times 10^{-8}$	W m^{-2}K^{-4}
molar gas constant	$8.314\,34$	J mol^{-1}K^{-1}
gravitational constant	$6.673\,2 \times 10^{-11}$	N m^2 Δkg^{-2}

1°	0.0174 5329 radians
1′	0.0002 9089 radians
1″	0.0000 0485 radians
1 radian	57.29578°
	57°.17′45″
π	3.1415 9265
$\log_{10}\pi$	0.4971 4987
e	2.7182 8183
$\log_{10}e$	0.4342 9448
$\log_{e}10$	2.3025 8509

Dimensions and Units of some Physical Quantities

Quantity	Dimension	Unit
mass	$[M]$	kg
length	$[L]$	m
time	$[T]$	s
area	$[L^2]$	m^2
volume	$[L^3]$	m^3
density	$[ML^{-3}]$	kg m^{-3}
acceleration	$[LT^{-2}]$	m s^{-2}
force	$[MLT^{-2}]$	N
pressure	$[ML^{-1}T^{-2}]$	Pa
momentum	$[MLT^{-1}]$	N s
pulsatance	$[T^{-1}]$	Hz

The Greek Alphabet

Letters		Name	Letters		Name
A	α	alpha	N	ν	nu
ß	β	beta	Ξ	ξ	xi
Γ	γ	gamma	Ο	ο	omicron
Δ	δ	delta	Π	π	pi
Ε	ε	epsilon	Ρ	ρ	rho
Ζ	ζ	zeta	Σ	σ	sigma
Η	η	eta	Τ	τ	tau
Θ	θ	theta	Τ	ν	upsilon
Ι	ι	iota	Φ	φ	phi
Κ	κ	kappa	Χ	χ	chi
Λ	λ	lambda	Ψ	ψ	psi
Μ	μ	mu	Ω	ω	omega